T0133170

Statistical Methods in Computer Security

Statistical Distributions
Professor N. Balakrishnan
McMaster University

Statistical Process Improvement
Professor G. Geoffrey Vining
Virginia Polytechnic Institute

Stochastic Processes
Professor V. Lakshmikantham
Florida Institute of Technology

Survey Sampling
Professor Lynne Stokes
Southern Methodist University

Time Series
Sastry G. Pantula
North Carolina State University

1. The Generalized Jackknife Statistic, *H. L. Gray and W. R. Schucany*
2. Multivariate Analysis, *Anant M. Kshirsagar*
3. Statistics and Society, *Walter T. Federer*
4. Multivariate Analysis: A Selected and Abstracted Bibliography, 1957–1972, *Kocherlakota Subrahmaniam and Kathleen Subrahmaniam*
5. Design of Experiments: A Realistic Approach, *Virgil L. Anderson and Robert A. McLean*
6. Statistical and Mathematical Aspects of Pollution Problems, *John W. Pratt*
7. Introduction to Probability and Statistics (in two parts), Part I: Probability; Part II: Statistics, *Narayan C. Giri*
8. Statistical Theory of the Analysis of Experimental Designs, *J. Ogawa*
9. Statistical Techniques in Simulation (in two parts), *Jack P. C. Kleijnen*
10. Data Quality Control and Editing, *Joseph I. Naus*
11. Cost of Living Index Numbers: Practice, Precision, and Theory, *Kali S. Banerjee*
12. Weighing Designs: For Chemistry, Medicine, Economics, Operations Research, Statistics, *Kali S. Banerjee*
13. The Search for Oil: Some Statistical Methods and Techniques, *edited by D. B. Owen*
14. Sample Size Choice: Charts for Experiments with Linear Models, *Robert E. Odeh and Martin Fox*
15. Statistical Methods for Engineers and Scientists, *Robert M. Bethea, Benjamin S. Duran, and Thomas L. Boullion*
16. Statistical Quality Control Methods, *Irving W. Burr*

17. On the History of Statistics and Probability, *edited by D. B. Owen*
18. Econometrics, *Peter Schmidt*
19. Sufficient Statistics: Selected Contributions, *Vasant S. Huzurbazar (edited by Anant M. Kshirsagar)*
20. Handbook of Statistical Distributions, *Jagdish K. Patel, C. H. Kapadia, and D. B. Owen*
21. Case Studies in Sample Design, *A. C. Rosander*
22. Pocket Book of Statistical Tables, *compiled by R. E. Odeh, D. B. Owen, Z. W. Birnbaum, and L. Fisher*
23. The Information in Contingency Tables, *D. V. Gokhale and Solomon Kullback*
24. Statistical Analysis of Reliability and Life-Testing Models: Theory and Methods, *Lee J. Bain*
25. Elementary Statistical Quality Control, *Irving W. Burr*
26. An Introduction to Probability and Statistics Using BASIC, *Richard A. Groeneveld*
27. Basic Applied Statistics, *B. L. Raktoe and J. J. Hubert*
28. A Primer in Probability, *Kathleen Subrahmaniam*
29. Random Processes: A First Look, *R. Syski*
30. Regression Methods: A Tool for Data Analysis, *Rudolf J. Freund and Paul D. Minton*
31. Randomization Tests, *Eugene S. Edgington*
32. Tables for Normal Tolerance Limits, Sampling Plans and Screening, *Robert E. Odeh and D. B. Owen*
33. Statistical Computing, *William J. Kennedy, Jr., and James E. Gentle*
34. Regression Analysis and Its Application: A Data-Oriented Approach, *Richard F. Gunst and Robert L. Mason*
35. Scientific Strategies to Save Your Life, *I. D. J. Bross*
36. Statistics in the Pharmaceutical Industry, *edited by C. Ralph Buncher and Jia-Yeong Tsay*
37. Sampling from a Finite Population, *J. Hajek*
38. Statistical Modeling Techniques, *S. S. Shapiro and A. J. Gross*
39. Statistical Theory and Inference in Research, *T. A. Bancroft and C.-P. Han*
40. Handbook of the Normal Distribution, *Jagdish K. Patel and Campbell B. Read*
41. Recent Advances in Regression Methods, *Hrishikesh D. Vinod and Aman Ullah*
42. Acceptance Sampling in Quality Control, *Edward G. Schilling*
43. The Randomized Clinical Trial and Therapeutic Decisions, *edited by Niels Tygstrup, John M Lachin, and Erik Juhl*
44. Regression Analysis of Survival Data in Cancer Chemotherapy, *Walter H. Carter, Jr., Galen L. Wampler, and Donald M. Stablein*
45. A Course in Linear Models, *Anant M. Kshirsagar*
46. Clinical Trials: Issues and Approaches, *edited by Stanley H. Shapiro and Thomas H. Louis*
47. Statistical Analysis of DNA Sequence Data, *edited by B. S. Weir*
48. Nonlinear Regression Modeling: A Unified Practical Approach, *David A. Ratkowsky*
49. Attribute Sampling Plans, Tables of Tests and Confidence Limits for Proportions, *Robert E. Odeh and D. B. Owen*

50. Experimental Design, Statistical Models, and Genetic Statistics, *edited by Klaus Hinkelmann*
51. Statistical Methods for Cancer Studies, *edited by Richard G. Cornell*
52. Practical Statistical Sampling for Auditors, *Arthur J. Wilburn*
53. Statistical Methods for Cancer Studies, *edited by Edward J. Wegman and James G. Smith*
54. Self-Organizing Methods in Modeling: GMDH Type Algorithms, *edited by Stanley J. Farlow*
55. Applied Factorial and Fractional Designs, *Robert A. McLean and Virgil L. Anderson*
56. Design of Experiments: Ranking and Selection, *edited by Thomas J. Santner and Ajit C. Tamhane*
57. Statistical Methods for Engineers and Scientists: Second Edition, Revised and Expanded, *Robert M. Bethea, Benjamin S. Duran, and Thomas L. Boullion*
58. Ensemble Modeling: Inference from Small-Scale Properties to Large-Scale Systems, *Alan E. Gelfand and Crayton C. Walker*
59. Computer Modeling for Business and Industry, *Bruce L. Bowerman and Richard T. O'Connell*
60. Bayesian Analysis of Linear Models, *Lyle D. Broemeling*
61. Methodological Issues for Health Care Surveys, *Brenda Cox and Steven Cohen*
62. Applied Regression Analysis and Experimental Design, *Richard J. Brook and Gregory C. Arnold*
63. Statpal: A Statistical Package for Microcomputers—PC-DOS Version for the IBM PC and Compatibles, *Bruce J. Chalmer and David G. Whitmore*
64. Statpal: A Statistical Package for Microcomputers—Apple Version for the II, II+, and IIe, *David G. Whitmore and Bruce J. Chalmer*
65. Nonparametric Statistical Inference: Second Edition, Revised and Expanded, *Jean Dickinson Gibbons*
66. Design and Analysis of Experiments, *Roger G. Petersen*
67. Statistical Methods for Pharmaceutical Research Planning, *Sten W. Bergman and John C. Gittins*
68. Goodness-of-Fit Techniques, *edited by Ralph B. D'Agostino and Michael A. Stephens*
69. Statistical Methods in Discrimination Litigation, *edited by D. H. Kaye and Mikel Aickin*
70. Truncated and Censored Samples from Normal Populations, *Helmut Schneider*
71. Robust Inference, *M. L. Tiku, W. Y. Tan, and N. Balakrishnan*
72. Statistical Image Processing and Graphics, *edited by Edward J. Wegman and Douglas J. DePriest*
73. Assignment Methods in Combinatorial Data Analysis, *Lawrence J. Hubert*
74. Econometrics and Structural Change, *Lyle D. Broemeling and Hiroki Tsurumi*
75. Multivariate Interpretation of Clinical Laboratory Data, *Adelin Albert and Eugene K. Harris*
76. Statistical Tools for Simulation Practitioners, *Jack P. C. Kleijnen*
77. Randomization Tests: Second Edition, *Eugene S. Edgington*

78. A Folio of Distributions: A Collection of Theoretical Quantile-Quantile Plots, *Edward B. Fowlkes*
79. Applied Categorical Data Analysis, *Daniel H. Freeman, Jr.*
80. Seemingly Unrelated Regression Equations Models: Estimation and Inference, *Virendra K. Srivastava and David E. A. Giles*
81. Response Surfaces: Designs and Analyses, *Andre I. Khuri and John A. Cornell*
82. Nonlinear Parameter Estimation: An Integrated System in BASIC, *John C. Nash and Mary Walker-Smith*
83. Cancer Modeling, *edited by James R. Thompson and Barry W. Brown*
84. Mixture Models: Inference and Applications to Clustering, *Geoffrey J. McLachlan and Kaye E. Basford*
85. Randomized Response: Theory and Techniques, *Arijit Chaudhuri and Rahul Mukerjee*
86. Biopharmaceutical Statistics for Drug Development, *edited by Karl E. Peace*
87. Parts per Million Values for Estimating Quality Levels, *Robert E. Odeh and D. B. Owen*
88. Lognormal Distributions: Theory and Applications, *edited by Edwin L. Crow and Kunio Shimizu*
89. Properties of Estimators for the Gamma Distribution, *K. O. Bowman and L. R. Shenton*
90. Spline Smoothing and Nonparametric Regression, *Randall L. Eubank*
91. Linear Least Squares Computations, *R. W. Farebrother*
92. Exploring Statistics, *Damaraju Raghavarao*
93. Applied Time Series Analysis for Business and Economic Forecasting, *Sufi M. Nazem*
94. Bayesian Analysis of Time Series and Dynamic Models, *edited by James C. Spall*
95. The Inverse Gaussian Distribution: Theory, Methodology, and Applications, *Raj S. Chhikara and J. Leroy Folks*
96. Parameter Estimation in Reliability and Life Span Models, *A. Clifford Cohen and Betty Jones Whitten*
97. Pooled Cross-Sectional and Time Series Data Analysis, *Terry E. Dielman*
98. Random Processes: A First Look, Second Edition, Revised and Expanded, *R. Syski*
99. Generalized Poisson Distributions: Properties and Applications, *P. C. Consul*
100. Nonlinear L_p-Norm Estimation, *Rene Gonin and Arthur H. Money*
101. Model Discrimination for Nonlinear Regression Models, *Dale S. Borowiak*
102. Applied Regression Analysis in Econometrics, *Howard E. Doran*
103. Continued Fractions in Statistical Applications, *K. O. Bowman and L. R. Shenton*
104. Statistical Methodology in the Pharmaceutical Sciences, *Donald A. Berry*
105. Experimental Design in Biotechnology, *Perry D. Haaland*
106. Statistical Issues in Drug Research and Development, *edited by Karl E. Peace*
107. Handbook of Nonlinear Regression Models, *David A. Ratkowsky*

108. Robust Regression: Analysis and Applications, *edited by Kenneth D. Lawrence and Jeffrey L. Arthur*
109. Statistical Design and Analysis of Industrial Experiments, *edited by Subir Ghosh*
110. *U*-Statistics: Theory and Practice, *A. J. Lee*
111. A Primer in Probability: Second Edition, Revised and Expanded, *Kathleen Subrahmaniam*
112. Data Quality Control: Theory and Pragmatics, *edited by Gunar E. Liepins and V. R. R. Uppuluri*
113. Engineering Quality by Design: Interpreting the Taguchi Approach, *Thomas B. Barker*
114. Survivorship Analysis for Clinical Studies, *Eugene K. Harris and Adelin Albert*
115. Statistical Analysis of Reliability and Life-Testing Models: Second Edition, *Lee J. Bain and Max Engelhardt*
116. Stochastic Models of Carcinogenesis, *Wai-Yuan Tan*
117. Statistics and Society: Data Collection and Interpretation, Second Edition, Revised and Expanded, *Walter T. Federer*
118. Handbook of Sequential Analysis, *B. K. Ghosh and P. K. Sen*
119. Truncated and Censored Samples: Theory and Applications, *A. Clifford Cohen*
120. Survey Sampling Principles, *E. K. Foreman*
121. Applied Engineering Statistics, *Robert M. Bethea and R. Russell Rhinehart*
122. Sample Size Choice: Charts for Experiments with Linear Models: Second Edition, *Robert E. Odeh and Martin Fox*
123. Handbook of the Logistic Distribution, *edited by N. Balakrishnan*
124. Fundamentals of Biostatistical Inference, *Chap T. Le*
125. Correspondence Analysis Handbook, *J.-P. Benzécri*
126. Quadratic Forms in Random Variables: Theory and Applications, *A. M. Mathai and Serge B. Provost*
127. Confidence Intervals on Variance Components, *Richard K. Burdick and Franklin A. Graybill*
128. Biopharmaceutical Sequential Statistical Applications, *edited by Karl E. Peace*
129. Item Response Theory: Parameter Estimation Techniques, *Frank B. Baker*
130. Survey Sampling: Theory and Methods, *Arijit Chaudhuri and Horst Stenger*
131. Nonparametric Statistical Inference: Third Edition, Revised and Expanded, *Jean Dickinson Gibbons and Subhabrata Chakraborti*
132. Bivariate Discrete Distribution, *Subrahmaniam Kocherlakota and Kathleen Kocherlakota*
133. Design and Analysis of Bioavailability and Bioequivalence Studies, *Shein-Chung Chow and Jen-pei Liu*
134. Multiple Comparisons, Selection, and Applications in Biometry, *edited by Fred M. Hoppe*
135. Cross-Over Experiments: Design, Analysis, and Application, *David A. Ratkowsky, Marc A. Evans, and J. Richard Alldredge*
136. Introduction to Probability and Statistics: Second Edition, Revised and Expanded, *Narayan C. Giri*

137. Applied Analysis of Variance in Behavioral Science, *edited by Lynne K. Edwards*
138. Drug Safety Assessment in Clinical Trials, *edited by Gene S. Gilbert*
139. Design of Experiments: A No-Name Approach, *Thomas J. Lorenzen and Virgil L. Anderson*
140. Statistics in the Pharmaceutical Industry: Second Edition, Revised and Expanded, *edited by C. Ralph Buncher and Jia-Yeong Tsay*
141. Advanced Linear Models: Theory and Applications, *Song-Gui Wang and Shein-Chung Chow*
142. Multistage Selection and Ranking Procedures: Second-Order Asymptotics, *Nitis Mukhopadhyay and Tumulesh K. S. Solanky*
143. Statistical Design and Analysis in Pharmaceutical Science: Validation, Process Controls, and Stability, *Shein-Chung Chow and Jen-pei Liu*
144. Statistical Methods for Engineers and Scientists: Third Edition, Revised and Expanded, *Robert M. Bethea, Benjamin S. Duran, and Thomas L. Boullion*
145. Growth Curves, *Anant M. Kshirsagar and William Boyce Smith*
146. Statistical Bases of Reference Values in Laboratory Medicine, *Eugene K. Harris and James C. Boyd*
147. Randomization Tests: Third Edition, Revised and Expanded, *Eugene S. Edgington*
148. Practical Sampling Techniques: Second Edition, Revised and Expanded, *Ranjan K. Som*
149. Multivariate Statistical Analysis, *Narayan C. Giri*
150. Handbook of the Normal Distribution: Second Edition, Revised and Expanded, *Jagdish K. Patel and Campbell B. Read*
151. Bayesian Biostatistics, *edited by Donald A. Berry and Dalene K. Stangl*
152. Response Surfaces: Designs and Analyses, Second Edition, Revised and Expanded, *André I. Khuri and John A. Cornell*
153. Statistics of Quality, *edited by Subir Ghosh, William R. Schucany, and William B. Smith*
154. Linear and Nonlinear Models for the Analysis of Repeated Measurements, *Edward F. Vonesh and Vernon M. Chinchilli*
155. Handbook of Applied Economic Statistics, *Aman Ullah and David E. A. Giles*
156. Improving Efficiency by Shrinkage: The James-Stein and Ridge Regression Estimators, *Marvin H. J. Gruber*
157. Nonparametric Regression and Spline Smoothing: Second Edition, *Randall L. Eubank*
158. Asymptotics, Nonparametrics, and Time Series, *edited by Subir Ghosh*
159. Multivariate Analysis, Design of Experiments, and Survey Sampling, *edited by Subir Ghosh*
160. Statistical Process Monitoring and Control, *edited by Sung H. Park and G. Geoffrey Vining*
161. Statistics for the 21st Century: Methodologies for Applications of the Future, *edited by C. R. Rao and Gábor J. Székely*
162. Probability and Statistical Inference, *Nitis Mukhopadhyay*

163. Handbook of Stochastic Analysis and Applications, *edited by D. Kannan and V. Lakshmikantham*
164. Testing for Normality, *Henry C. Thode, Jr.*
165. Handbook of Applied Econometrics and Statistical Inference, *edited by Aman Ullah, Alan T. K. Wan, and Anoop Chaturvedi*
166. Visualizing Statistical Models and Concepts, *R. W. Farebrother and Michael Schyns*
167. Financial and Actuarial Statistics, *Dale Borowiak*
168. Nonparametric Statistical Inference, Fourth Edition, Revised and Expanded, *edited by Jean Dickinson Gibbons and Subhabrata Chakraborti*
169. Computer-Aided Econometrics, *edited by David EA. Giles*
170. The EM Algorithm and Related Statistical Models, *edited by Michiko Watanabe and Kazunori Yamaguchi*
171. Multivariate Statistical Analysis, Second Edition, Revised and Expanded, *Narayan C. Giri*
172. Computational Methods in Statistics and Econometrics, *Hisashi Tanizaki*
173. Applied Sequential Methodologies: Real-World Examples with Data Analysis, *edited by Nitis Mukhopadhyay, Sujay Datta, and Saibal Chattopadhyay*
174. Handbook of Beta Distribution and Its Applications, *edited by Richard Guarino and Saralees Nadarajah*
175. Survey Sampling: Theory and Methods, Second Edition, *Arijit Chardhuri and Horst Stenger*
176. Item Response Theory: Parameter Estimation Techniques, Second Edition, *edited by Frank B. Baker and Seock-Ho Kim*
177. Statistical Methods in Computer Security, *edited by William W. S. Chen*
178. Elementary Statistical Quality Control, Second Edition, *John T. Burr*
179. Data Analysis of Asymmetric Structures, *Takayuki Saito and Hiroshi Yadohisa*

Additional Volumes in Preparation

Statistical Methods in Computer Security

edited by
William W. S. Chen
Internal Revenue Service
Washington, D.C., U.S.A.

MARCEL DEKKER NEW YORK

Although great care has been taken to provide accurate and current information, neither the author(s) nor the publisher, nor anyone else associated with this publication, shall be liable for any loss, damage, or liability directly or indirectly caused or alleged to be caused by this book. The material contained herein is not intended to provide specific advice or recommendations for any specific situation.

Library of Congress Cataloging-in-Publication Data
A catalog record for this book is available from the Library of Congress.

ISBN: 0-8247-5939-7

This book is printed on acid-free paper.

Headquarters
Marcel Dekker, 270 Madison Avenue, New York, NY 10016, U.S.A.
tel: 212-696-9000; fax: 212-685-4540

Distribution and Customer Service
Marcel Dekker, Cimarron Road, Monticello, New York 12701, U.S.A.
tel: 800-228-1160; fax: 845-796-1772

World Wide Web
http://www.dekker.com

The publisher offers discounts on this book when ordered in bulk quantities. For more information, write to Special Sales/Professional Marketing at the headquarters address above.

Current printing (last digit):

10 9 8 7 6 5 4 3 2 1

PRINTED IN THE UNITED STATES OF AMERICA

Preface

Computer security is a global challenge for the new century, and has received a great deal of attention in recent years. This anthology explores security methods from a variety of viewpoints; from academia, industry, and government. The main focus covers five major areas: computer security policy, internet security, firewalls, virus protection, and statistical methods in computer security. The contents bring together the importance of both theory and application, and should interest researchers and practitioners alike.

The area of computer security management is vast, and impossible to cover in a single volume. We made every effort to include papers that are accessible to the general reader but also have sufficient depth for computer security analysts. This was done with the hope that the majority of readers would be able to more easily digest and use the information presented in the various sections.

Statistical Methods In Computer Security would not have been possible without the ongoing support of all the contributing authors. Some of the papers were reorganized, retyped and sent back to authors for review multiple times.

The authors had to go through the process of submitting the final manuscripts, responding to comments and then working together to produce a publishable volume. I sincerely appreciate the authors' diligent efforts. Thanks are also due to Marcel Dekker Publishers for fully supporting the effort to bring this book to fruition.

Willaim W. Chen
Internal Revenue Service
March 2004

Contents

Contributors

M. L. Adams Naval Surface Warfare Center, Dahlgren, Virginia, USA

James P. Anderson James P. Anderson Company

Matt Bishop Department of Computer Science, University of California at Davis, Davis, California, USA

Sheila Brand National Security Agency, US Department of Defense

Thomas M. Chen Department of Electrical Engineering, Southern Methodist University, Dallas, Texas, USA

Steven Cheung Department of Computer Science, University of California at Davis, Davis, California, USA

Paul C. Clark SecureMethods, Inc., Vienna, Virginia, USA

George Cybenko Institute for Security Technology Studies, Thayer School of Engineering, Dartmouth College, Hanover, New Hampshire, USA

Jeremy Epstein Product Security & Performance, webMethods, Inc., Fairfax, Virginia, USA

Jeremy Frank Department of Computer Science, University of California at Davis, Davis, California, USA

Li Gong JavaSoft, Sun Microsystems

Sumit Gupta Sun Microsystems Laboratories, Mountain View, California, USA

Vipul Gupta Sun Microsystems Laboratories, Mountain View, California, USA

Jeff Hayes CISSP, Cerberian, Inc., Draper, Utah, USA

James Hoagland Department of Computer Science, University of California at Davis, Davis, California, USA

Guofei Jiang Institute for Security Technology Studies, Thayer School of Engineering, Dartmouth College, Hanover, New Hampshire, USA

Richard J. Kryscio Department of Statistics, University of Kentucky, Lexington, Kentucky, USA

Claude Lefevre Institut de Statistique et de Recherche Operationelle, Université Libre de Bruxelles, Brussels, Belgium

Sharman Lichtenstein School of Information Systems, Deakin University, Melbourne, Australia

David J. Marchette Naval Surface Warfare Center, Dahlgren, Virginia, USA

Tom Markham Secure Computing Corporation, Roseville, California, USA

Gabriel Mateescu Research Computing Support Group, National Research Council Canada, Ottawa, Ontario, Canada

Dennis McGrath Institute for Security Technology Studies, Thayer School of Engineering, Dartmouth College, Hanover, New Hampshire, USA

Gary McGraw Cigital, Dulles, Virginia, USA

Catherine Meadows Naval Research Laboratory, Washington, DC, USA

Marion C. Meissner SecureMethods, Inc., Vienna, Virginia, USA

Eric Monteith McAfee Labs, Herdon, Virginia, USA

Charles Payne Secure Computing Corporation, Roseville, California, USA

Eddie Rabinovitch ECI Technology, East Rutherford, New Jersey, USA

Jean-Marc Robert Alcatel Canada Inc., Ottawa, Ontario, Canada

Steven Samorodin Department of Computer Science, University of California at Davis, Davis, California, USA

J. L. Solka Naval Surface Warfare Center, Dahlgren, Virginia, USA

Masha Sosonkina Scalable Computing Laboratory, Ames Laboratory, Ames, Iowa, USA

Linda Thomas Product Security & Performance, webMethods, Inc., Fairfax, Virginia, USA

Paul Thompson Thayer School of Engineering, Dartmouth College, Hanover, New Hampshire, USA

Karen O. Vance SecureMethods, Inc., Vienna, Virginia, USA

Dennis Volpano Computer Science Department, Naval Postgraduate School, Monterey, California, USA

Chris Wee Department of Computer Science, University of California at Davis, Davis, California, USA

E. J. Wegman Center for Computational Statistics, George Mason University, Fairfax, Virginia, USA

John C. Wierman Department of Applied Mathematics and Statistics, Johns Hopkins University, Baltimore, Maryland, USA

1

Development of Organizational Internet Security Policy: A Holistic Approach

SHARMAN LICHTENSTEIN
School of Information Systems,
Deakin University, Melbourne,
Australia

ABSTRACT

This chapter describes a general framework for developing organizational internet security policy. A model of internet security risks for an internet user organization is proposed. The framework utilizes this model, as well as a holistic approach, to develop the organization's internet security policy. A hierarchy of sub-policies for the internet security policy is also suggested. This chapter presents findings from part of a wider investigation into internet security policy.

1. INTRODUCTION

Although organizations clearly desire to encourage e-business and open access across the internet, they also wish to protect information resources, corporate images, the rights and needs of internal internet users, and salient interests of external parties. Companies also strive to remain law-abiding, and avoid liability from internet usage. To achieve these objectives, organizations seek a high level of internet security, defined as:

> the protection of the confidentiality, integrity and availability of the organization's information resources, and the protection of the organization's image, reputation, finance and viability, from accidental misuse or deliberate attack via internet connectivity.
>
> (Lichtenstein and Swatman, 1997)

Since its beginning in the 1970s, the internet has exhibited multifarious vulnerabilities—in its underlying communications network and nodes, internet protocols, network administration and host systems (Ghosh, 1998; Greenstein and Vasarhelyi, 2001). The increased vulnerability of the internet infrastructure to external attack has been well remarked, and has been a key finding in surveys of computer crime and security issues in organizations. For example, in a CSI/FBI survey of computer crime and security issues in American companies, 74% of respondents reported the internet as a source of frequent attack over the prior 12 months (CSI, 2002), with attacks having steadily increased over the previous six years of the survey.

Hackers, competitors, spies, disgruntled employees, and ex-employees continue to exploit the internet's ever-changing vulnerabilities, resulting in damage, disruption, and uncertainty—a sad indication of the changed nature of the internet environment from collegial and trustworthy, to competitive and hostile.

Particularly disturbing for companies, many employees who have been granted connection to the internet for valid business reasons have been misusing or abusing it (CSI, 2002), from a lack of awareness of the internet's insecurities, a lack of awareness of valid, value-adding business internet usages, or malicious intent.

In recent years, well-publicized security incidents have raised public awareness regarding the internet's vulnerabilities. Distributed denial-of-service attacks occurring at Yahoo!, Buy.com, eBay and E*Trade Web sites in February, 2000, have served as a warning of impending higher-impact attacks. The penetration of high profile US institutions, including the Department of Defense, NASA, the Pentagon, CIA and Citibank, highlighted the seriousness of the hacking threat. Costly virus incidents have been afforded significant media attention, illustrated by the Love Bug virus—which affected some 45 million users and caused an estimated $4 billion in damages, and the Melissa virus—which caused an estimated $80 million in damages in 1999. Accounts of internet credit card fraud abound—for example, the recent case of Adil Shakour, who hacked into various company databases to extract details of credit-card holders, carried out fraudulent purchases, and caused an estimated $100,000 in damage (DOJ, 2003). Neuman (2000) estimated losses from internet credit-card fraud at around $1 billion per annum.

In this way, organizations have become aware of the significance of the internet security problem, instigating in response a range of piecemeal solutions, such as firewalls. Clearly, management of internet security issues at a number of levels is required (Pethia et al., 2000), including at the organizational level, which is the focus of this paper. Policies, procedures, standards, and other management instructions are considered critical in the management of organizational internet security issues such as employee misuse and abuse of the internet, as they are aimed at controlling the decisive human factor—the company's employees (Bernstein et al., 1996; Guttman and Bagwill, 1997; Overly, 1999).

There is an important need for an organization to specify an internet security policy describing the requirements for internet security within the organization. Mechanisms which implement these particular requirements can then be selected. These policies make employees responsible and accountable for understanding and following security rules, and ensure that employees use available mechanisms to protect their systems. The internet security policy should be a key component of an overarching internet security program in the organization. Pethia et al. (1991) wrote:

> There must be a clear statement of the local (internet) security policy, and this policy must be communicated to the users and other relevant parties. The policy should be on file and available to users at all times, and should be communicated to users as part of providing access to the system.

Until recently, there was a dearth of internet security management guidelines for organizations. Early efforts in producing guidelines for internet security policy included Pethia et al. (1991), Bernstein et al. (1996), and NIST guidelines of Guttman and Bagwill (1997). More recently, as part of a comprehensive four year research project conducted from 1996 to 2000, Lichtenstein (2001) developed a set of guidelines for internet security policy, with the work reported in this chapter presenting selected findings from that project.

Over the past decade, a number of researchers have argued for a holistic perspective in both information security and internet security for example, Brunnstein, 1997; Yngstrom, 1995; Lichtenstein, 2001; Lichtenstein and Swatman, 2001), suggesting that guidelines in internet security policy should accommodate this perspective.

A method is required to develop an internet security policy. The method should feature a holistic approach, accounting for a diverse range of issues including internet risks, and resulting in a well-structured, comprehensive and effective internet security policy (Lichtenstein, 2001; Lichtenstein and Swatman, 2001). This chapter addresses the question:

Can a framework for developing an organizational internet security policy be specified, which adopts an holistic approach?

The chapter reports results from one part of a wider investigation into internet security policy (described in detail in Lichtenstein, 2001), and provides a significant update, now that the project has been completed, of earlier work reported in Lichtenstein (1997). The wider investigation involved subjective/argumentative research followed by empirical work consisting of six case studies of large organizations, and validation of research findings by a focus group of experts (Lichtenstein, 2001).

The chapter begins by describing a high-level internet risk model for an organization. This is followed by a description and discussion of a holistic model of the factors influencing internet security policy. A framework is then described for the development of organizational internet security policy. The chapter also provides a hierarchy of sub-policies to guide the content of internet security policy. Finally, findings are discussed and conclusions drawn.

2. ORGANIZATIONAL INTERNET RISKS

In the development of an internet security policy, it is imperative that the risks to the organization arising from the internet connection be addressed. This requires identification of the relevant risks, followed by risk prioritization via a risk assessment process. The research project identified a set of internet risk types, illustrated in Figure 1.

Much has been written about organizational internet risks (for example, Greenstein and Vasarhelyi, 2001). It is beyond the charter of this paper to identify or discuss in depth the many and varied risks; however, the above source as well as others may be consulted in order to add richness to the internet risks model illustrated in Figure 1.

In the model, risks are categorized into risk types, in order to assist identification of existing internet risks for the organization. The impacts of the risk types are loss of confidentiality, integrity or availability to either the organization's information resources or to other internet participants' information resources, or damage to the organization's image, reputation, finances or viability. The model also includes risks

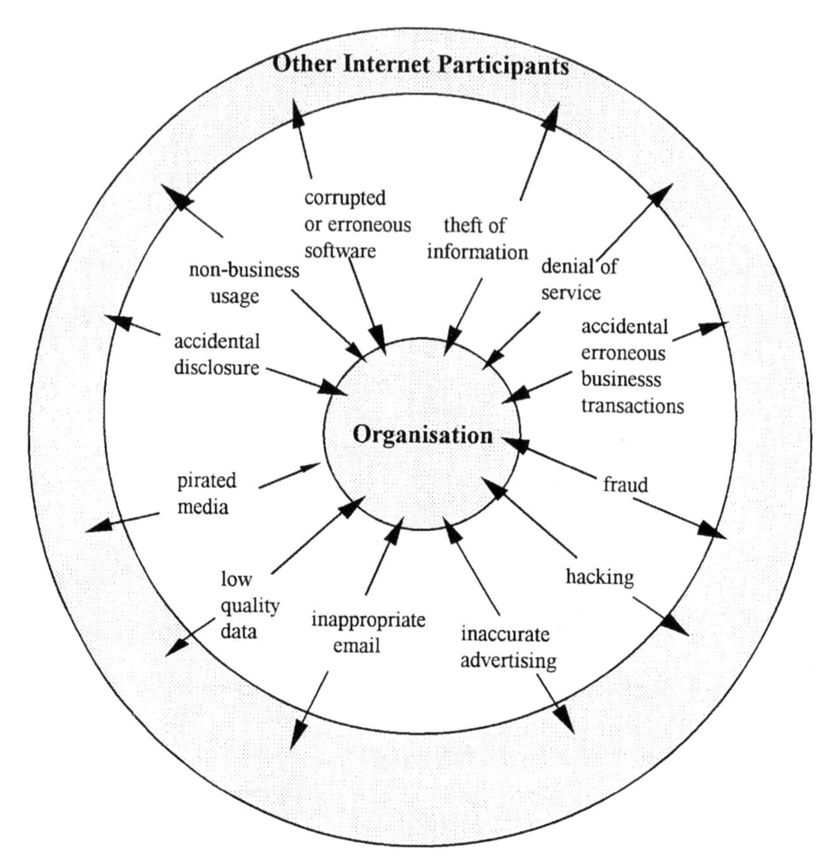

Figure 1 Internet risks for an organization.

which infringe upon employee privacy rights (for example, the sending or receipt of inappropriate email by employees).

The central circle denotes an organization with internet connection. The outer ring labeled "Other Internet Participants" denotes other members of the internet community with whom the organization communicates via the internet. The arrows portray internet risks which can emanate from within the organization and affect other internet participants, or which can emanate from other internet participants and affect the organization.

2.1. Accidental Erroneous Business Transactions

Electronic messages may become corrupted accidentally in transit (for instance, corrupted EDI messages), or companies may accidentally issue transactions incorrectly—a well-known example being the misdirection of email.

2.2. Low Quality Data

The quality of data being exchanged via the internet is questionable, in that it may be inaccurate, untimely, inconsistent, or merely opinion rather than fact (Mathieu and Woodard, 1995). Organizations or employees may accidentally issue transactions which contain incorrect data (for example, by inclusion of an inaccurate data field).

Another manifestation of the problem is that initially correct information, in the form of business transaction data or communications, may become altered in transit, either deliberately, via eavesdropping (also known as sniffing or snooping), or accidentally (Mathieu and Woodard, 1995). Outdated (i.e. untimely) information may remain on old Web sites. Conflicting versions of data may exist—for example, two versions of a database. Subjective opinion, rather than fact, may be transmitted by organizations or individuals, via postings.

2.3. Non-business Usage

Employees may be using the internet for a variety of non-business activities, including personal surfing, chatting, message boards, downloading games and images, personal e-mail, and personal use of other tools (for example, video-conferencing, hobby groups and newsgroups).

2.4. Accidental Disclosure

Employees may be incautious in their use of the internet when communicating possibly confidential business matters. An example of accidental disclosure is the inclusion of confidential information within email, a Web site, or another posting mechanism.

2.5. Denial of Service

Internet participants may act as a source or conduit for deliberate threats resulting in denial of service, either manifested as a internet traffic delay or total inability to gain access to the internet. Alternately, resource over-usage may accidentally slow or disable the internet facility.

2.6 Fraud

Internet participants may fraudulently obtain monies from a company or individuals using the internet connection, or may deliberately mislead other parties to secure advantages (for example, scams).

2.7. Inappropriate Email

Companies or employees may send or receive unwanted or unsolicited email (junk email), harassing, discriminatory or defamatory email (flame email) or excessive, unwanted email (spamming).

2.8. Theft of Information

Information may be accessed and stolen from an organization via a variety of means, such as hacking into systems and viewing corporate data, eavesdropping at network nodes, copying Web site information without gaining owner permission, and downloading software illicitly (also, refer to the risk type 'pirated media').

2.9. Inaccurate Advertising

An organization or employee may 'advertise' within email, Web sites, or other posting mechanisms, in such a way as to appear to represent an official view. The content of this information may be inaccurate in an organizational context.

2.10. Hacking

An employee may gain unauthorized access to an organization's systems or data either out of curiosity or for a more harmful reason, and may subsequently cause damage. The well-known risk of impersonation is included in this risk type, two pertinent examples being the forging of email and the existence of undependable internet identifications.

2.11. Pirated Media

An employee may download software or data in breach of copyright or licensing laws.

2.12. Malicious Code and Erroneous Software

An employee may download software containing bugs, or malicious software such as viruses and trojan horse programs. Web browsers are particularly dangerous, in their provision of access to untrustworthy systems and in their invocation of unproven applications.

In all companies studied, all the abovementioned types of risks were routinely occurring, to varying degrees of risk. In particular, high level of non-business use and malicious code (viruses) were found, consistent with media publicity in recent years. Of note, in most of the companies studied, employees had been dismissed for significant personal internet use. Viruses were regularly encountered, with significant penetration occurring (for example, the Melissa virus of 1999). Clearly, internet risks are impacting e-business, necessitating improved management through internet security policy and other measures.

An internet security policy should be developed within the holistic context introduced earlier in the chapter, and discussed in detail below.

3. HOLISTIC FACTORS IN INTERNET SECURITY POLICY

The popular interpretation of holism is a study of the broad, all-encompassing picture, rather than a consideration of individual components alone. There has been considerable discussion by experts of holistic perspectives of information security. With information security being a weak-link phenomenon, its design needs to be multidimensional (NRC, 1991), addressing a broad range of issues including computer security, systems analysis and design methods, manual information systems, managerial information security issues (for example, security policies) and societal and ethical issues (Baskerville, 1988). There is a common theme to all holistic perspectives of information security—the non-technical issues should be considered equally important to the technical issues.

Examples of holistic information security perspectives include:

- The OECD's (1992) information security guidelines, which relate to many diverse aspects: people, their rights and responsibilities; viewpoints (multidisciplinary, inter-organizational, and intra-organizational); technical, administrative, organizational, operational, commercial, educational and legal aspects; the cooperation of parties; the integration of the parts to form a coherent information security system; and
- Organizational information security policies, which should take into account the organization's information security philosophy, national policy, international standards, political issues, relevant organizational policies, implementation platform limitations, and relevant ethical, legal, and privacy issues (Olson and Abrams, 1995).

While many holistic views of information security currently exist, holistic approaches are still required for developing, evaluating, and managing information security (Hartmann, 1995; Rannenberg, 1994; Yngstrom, 1995). Holistic perspectives are also being recommended for specific domains of information security, with this chapter advocating a holistic perspective for the domain of internet security. In the past, there has been a tendency to focus almost exclusively on technical security needs in internet security policy—for example, the availability of specific security technologies, such as intrusion detection software. The companies studied possessed many non-technical security needs, for example the requirement for legality of e-business security policy—that is, compliance with existing relevant laws such as copyright. It was clear that companies were unable to ignore sensitive non-technical issues, such as employee rights to privacy in internet usage. Overall, there must be reasonable accommodation of both technical and non-technical factors in internet security policy, suggesting that a holistic methodology is required for policy development—an approach where the organizational, contextual and human issues are given equal consideration to the technical issues.

The study identified a set of holistic factors to consider when establishing internet security policy: internet risks, organizational, administrative, legal, societal, technical, standards, and human issues (Figure 2):

- Internet risks must be addressed by the policy, as discussed earlier.
- Organizational and administrative issues can affect the policy. For example, those usages of the internet which are considered valid by the company, and therefore acceptable uses, should be articulated in the policy. These usages should be aligned with organizational objectives in order to maximize benefits to the company. Administrative and opera-

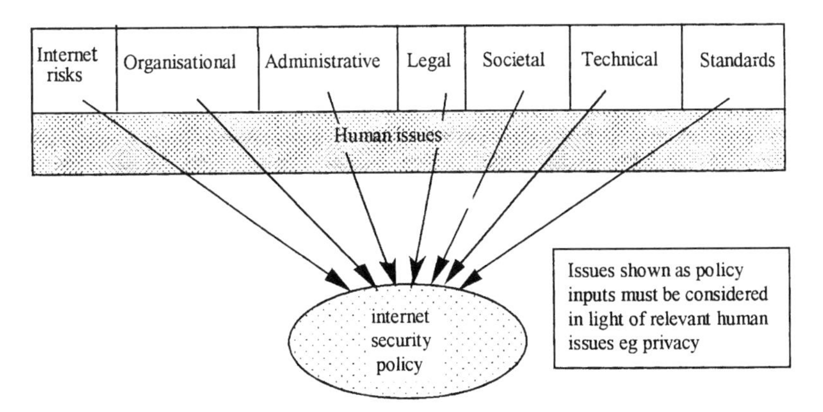

Figure 2 Holistic factors in internet security policy.

tional tasks must be defined—for example, procedures for applying, monitoring, and auditing policies.

- Legal issues must be considered when setting policy (for example, Saunders Thomas et al., 1998). The policy must reflect current laws (such as copyright laws), while the internet acceptable use policy informs employees of illegal internet actions (as well as other organizationally unacceptable internet uses).
- Societal issues should be considered in situations where the company perceives that the larger, global society is affected by its internet security management. For example, setting netiquette standards can take into account a company's interactions with other cultures.
- Technical issues include technical constraints that affect the outcome of the internet security policy. For example, available workstation technology places constraints on those internet services which the workstations will reasonably be able to utilize. The company must also consider likely expenditure on additional technologies to improve security, and formulate a policy which foreshadows the acquisition of these technologies (D'Alotto, 1996). If there has already been an investment made, for example in a firewall, the company can devise a policy to use this investment to mitigate internet risks.
- Standards must be considered, as policy must adhere to relevant industry, national, and other standards.
- Human issues affecting internal employees play a critical role in internet security policy. Indeed, an important finding from the research was that the human issues should be the filter through which all other factors are viewed, during policy development, as shown in Figure 2. The human issues identified in the study comprised: freedom of internet use, privacy, trust, monitoring, surveillance, censorship, right to be kept informed, accountability, sanctions, ownership, and ethics.

Employees will demand certain internet rights, and accept certain internet responsibilities and accountability, depending on the ethical, social, and cultural climate of the organization and nation. For example, in many organizations, employees may feel unduly restricted if prohibited entirely from utilizing the internet for personal reasons. They may feel they have the right to liberally send personal email, for example. Employees in many environments may refuse to accept accountability without adequate internet awareness programs in place, including provision of written policy which clearly defines both their responsibilities and the acceptable employee usages of the internet. Employees will need to have policy concepts explained clearly via security awareness sessions. Sanctions for breaching policy should be clearly defined, and acceptable. Employee privacy issues must also be addressed. For example, individual employees may not wish personal information about them to be published and made available on the world wide web. Other ethical concerns include Rannenberg's (1994) multilateral security concerns: unobservability, anonymity, unlinkability, pseudonymity and non-repudiation.

In all cases studied, the human issues dominated, with the employee presenting the greatest threat to a company's internet security armoire (Lichtenstein, 2001).

4. FRAMEWORK FOR DEVELOPING ORGANIZATIONAL INTERNET SECURITY POLICY

An information security program for an organization consists of various policies and procedures, security education, security management, and a range of security mechanisms. A typical strategy for the engineering of information security comprises four phases (Abrams et al., 1995)—a requirements definition phase, culminating in a corporate information security policy containing layers of policies and procedures; a design phase, resulting in a set of security mechanisms which implement the requirements; an integration phase, which results in the coordinated security system being put in place; and a certification or accreditation phase, which results in a certificate of accreditation being produced, if relevant.

The corporate information security policy document is of critical importance to an organization's information security program. It contains the complete information security requirements for the organization, in the form of layers of policies representing progressively more refined and progressively more rule-like policies, addressing different audiences and different aspects of information security. The creation of appropriate policies involves many choices and decisions, from high-level decisions concerning organizational objectives down to lower-level decisions regarding hardware. The internet security policy is a sub-policy of the corporate information security policy, and is therefore determined during the requirements definition phase.

A framework in which to develop an organization's internet security policy is illustrated in Figure 3. Note that the framework is based on risk assessment, a process nowadays consistently recommended for security policy decision-making (Bernstein et al., 1996; GAO, 1998; Guttman and Bagwill, 1997; ISO, 2000; Stanley, 1997). The framework shows the internet risks model (Figure 1) in the top left-hand corner. Company-specific internet data are shown in the top right-hand corner. The internet risks model is used in conjunction with the company-specific internet security data as input into a risk assessment process, in order to identify and prioritize significant internet risks.

These risks are then considered, in conjunction with other influential factors in internet security policy (Figure 2), in order to construct the internet security policy, consistent with the content of other relevant company policies such as Code of Ethics, and according to the structure suggested by an internet security policy content model. The internet security policy is clearly positioned within the corporate information security policy.

The risk assessment process considers internal and external internet risks, risks occurring at any of the company

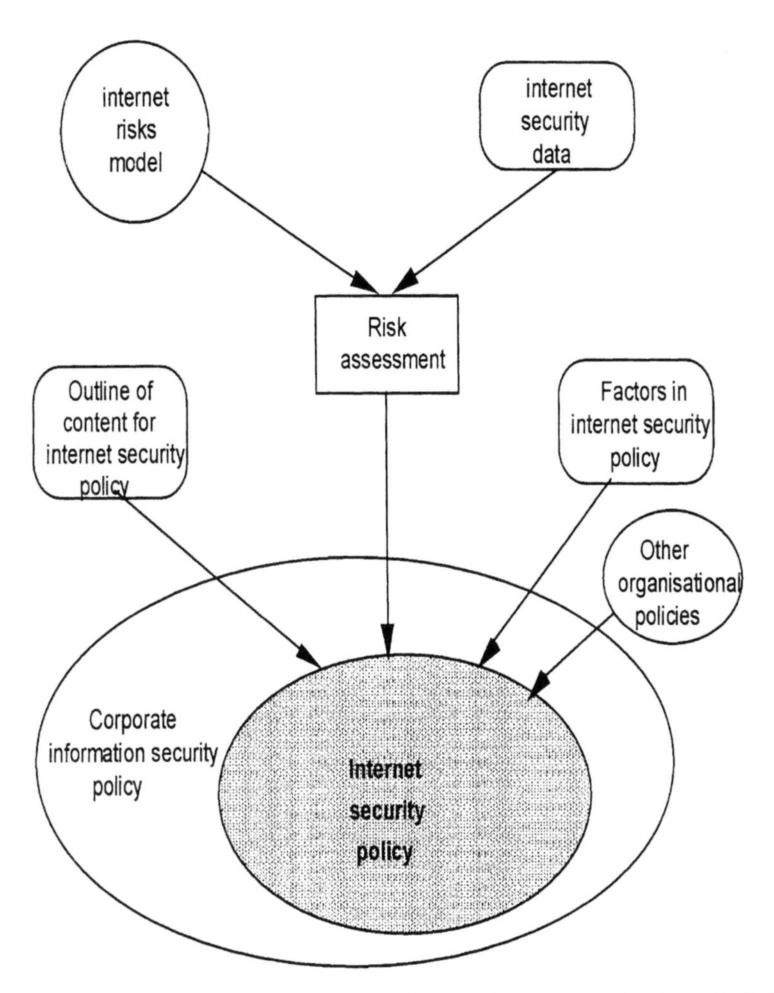

Figure 3 Framework for developing organizational internet security policy.

network's external access points, and the level of sensitivity and value of corporate data at risk. Calculations of risk for each identified internet risk and impacted resource can be performed using estimates of likelihood and impact, security statistics, professional opinion or other methods (Bernstein et al., 1996; GAO, 1998; Guttman and Bagwill, 1997). Resources which may be impacted include corporate data, internet hardware and software, and internet services. Risk likelihood and risk impact can be very difficult to quantitatively estimate, given the current lack of reliable metrics in this area. Therefore a qualitative risk assessment may be preferred, with each risk being attributed a qualifier such as high, medium or low. Guttman and Bagwill (1997) offer a method for qualitative internet risk assessment.

Useful characteristics to be developed in the internet security policy include flexibility, pertinence, applicability, implementability, timeliness, cost-effectiveness, enforceability, and integratibility (Bernstein et al., 1996).

5. INTERNET SECURITY POLICY CONTENT MODEL

Figure 3 utilizes an outline of the content of the internet security policy, to guide the development of the policy. The Internet Engineering Task Force (Pethia et al., 1991) specified six basic guidelines for internet security policies for internet user communities:

- assure individual accountability;
- employ available security mechanisms;
- maintain security of host computers;
- provide computers that embody security controls;
- cooperate in providing security; and
- seek technical improvements.

With its deeper empirical investigations, the research project which formed the basis of this chapter yielded an outline of the content for an internet security policy, comprised of a set of sub-policies, summarized in Table 1. Full details of this model are beyond scope of this chapter, and may be found in Lichtenstein (2001). The model can be employed by the framework in Figure 3 in policy development.

6. RESEARCH FINDINGS AND CONCLUSION

This chapter has focused on the role of organizational internet security policy in e-business security, providing a holistic

Table 1 Internet security policy content model

Purpose and scope
Philosophy
Organizational internet security infrastructure
Internet security management programme
Other applicable policies
Online privacy policy
Censorship policy
Accountability policy
Information protection policy
Information access policy
Firewall policy
Internet security technology policy
Password policy
Internet acceptable use policy
Internet publication policy
Email policy
Internet virus policy
Internet audit policy
Internet incident policy
Internet legal policy
Internet security policy review policy

approach to the development of the policy. Models were provided for internet risks, holistic factors influencing policy, policy development, and content outline. In all organizations studied during the project, a holistic approach had not been taken in developing various types of internet security policies, significantly contributing to the ineffectiveness of existing policies. There were significant internet risks occurring at each company studied, with non-business use and viruses the greatest concerns. Amongst many diverse factors identified as influential for internet security policy, the human issues dominated—including freedom of internet use, privacy, trust, monitoring, surveillance, censorship, right to be kept informed, accountability, sanctions, ownership, and ethics. It was recommended, therefore, that the human issues be the lens through which all other factors are viewed, when establishing various sub-policies of the internet security policy.

The study further identified disorganized management of internet security issues, at the companies studied. Internet security management was typically treated as an informal aspect of information security management, rather than in its own right—and, as a result, significant internet security issues were frequently overlooked. It was further apparent that none of the companies had taken the time to think through the various internet security issues, perhaps due to the absence of an organizational internet security infrastructure which would focus management attention. Finally, the study found widespread shortages of resources allocated to internet security management—a situation clearly requiring attention.

As a solution, therefore, the study supported the need for a well-resourced, formal organizational internet security infrastructure, featuring an internet security management program, centered on the internet security policy. The study also provided compelling support for viewing employees as key components of secure internet systems. Businesses are therefore advised to adopt a multifaceted approach to controlling the employee contribution to internet security concerns, comprising:

- the development of very secure systems;
- paying attention to the important human issues associated with internet security and usage;
- making employees accountable for their actions through appropriate policies, awareness activities, monitoring and sanctions; and
- minimizing reliance on employee behavior and actions through the development of powerful, emerging internet security management software.

In conclusion, with the looming threat of cyberterrorism in our changed global environment, the challenge now is to develop internet security solutions which will afford corporations the high level of protection needed to withstand prolonged and diverse attack. Only solutions based on strong, comprehensive, holistic internet security management and policy—such as the approach which has been outlined and argued in this chapter—will do.

REFERENCES

Abrams, M. D., Bailey, D. (1995). Abstraction and refinement of layered security policy. In: Abrams, M. D., Jajodia, S., Podell, H. J., eds. *Information Security—an Integrated Collection of Essays*. Los Alamitos, CA: IEEE Computer Society Press.

Abrams, M. D., Podell, H. J., Gambel, D. W. (1995). Security engineering. In: Abrams, M. D., Jajodia, S., Podell, H. J., eds. *Information Security—an Integrated Collection of Essays*. Los Alamitos, CA: IEEE Computer Society Press.

Baskerville, R. (1988). *Designing Information Systems Security*. John Wiley.

Bernstein, T., Bhimani, A. B., Schultz, E., Siegel, C. A. (1996). *Internet Security for Business*. John Wiley & Sons, Inc.

Brunnstein, K. (1997). Towards a holistic view of enterprise information and communication technologies: adapting to a changing paradigm. In: Yngstrom, Y., Carlsen, J., eds. *Information Security in Research and Business*. Proceedings of the 13th International Conference on Information Security (SEC '97), IFIP. Denmark: Chapman & Hall.

CSI (Computer Security Institute) (2002). *2002 Computer Crime and Security Survey*. San Francisco, USA: CSI.

D'Alotto, L. J. (1996). Internet firewalls policy development and technology choices. In: Proceedings of the 19th National Information Systems Security Conference, Baltimore, MD, US.

DOJ (Department of Justice) (2003). http://www.cybercrime.gov/shakourPlea.htm

GAO (1998) *Executive Guide Information Security Management*. USA: US General Accounting Office, May.

Ghosh, A. K. (1998). *E-Commerce Security: Weak Links, Best Defenses*. John Wiley & Sons.

Greenstein, M., Vasarhelyi, M. (2000). *Electronic Commerce: Security, Risk Management, and Control*. McGraw-Hill/Irwin.

Guttman, B., Bagwill, R. (1997) *Internet Security Policy: a Technical Guide—Draft*. Gaithersburg, MD: NIST Special Publication 800–XX.

Hartmann, A. (1995). Comprehensive information technology security: a new approach to respond ethical and social issues surrounding information security in the 21st century. In: *Information Security—the Next Decade*. IFIP/Sec '95, Proceedings of the IFIP TC11 11th International Conference on Information Security. Chapman Hall.

ISO (2000). *International Standard ISO/IEC 17799:2000 Code of Practice for Information Security Management*. Geneva, Switzerland: International Organization for Standardization.

Kaspersen, H. (1992). Security measures, standardisation and the law. In: Aiken, R., ed. Proceedings of the IFIP 12th World Computer Congress.

Lichtenstein, S. (1997). Developing Internet security policy for organizations. In: Nunamaker, J. F. Jr., Sprague, R. H. Jr., eds. Proceedings of the 13th Annual Hawaii International Conference on System Sciences (HICSS '97). Hawaii: IEEE Computer Society Press.

Lichtenstein, S. (1998). Internet risks for companies. *Computers & Security* 17(2).

Lichtenstein, S. (2001). Internet Security Policy for Organisations. Thesis (Ph.D.) (public version), School of Information Management and Systems, Monash University, Melbourne, Australia.

Lichtenstein, S., Swatman, P. M. C. (1997). Internet acceptable usage policy for organizations. *Information Management and Computer Security*. Vol. 5. No. 5. UK: *MCB University Press*.

Lichtenstein, S., Swatman, P. M. C. (2001). Effective management and policy in e-business security. In: Proceedings of the 14th Bled Electronic Commerce Conference, Bled, Slovenia.

Mathieu, R. G., Woodard, R. L. (1995). Data integrity and the internet: implications for management. *Information Management & Computer Security 3(2).*

Neumann, P. (2000). Note associated with "Hacking credit cards is preposterously easy". *Risks-forum Digest* 20(85).

NRC (1991). *Computers at Risk. Safe Computing in the Information Age, System Security Study Committee, Computer Science and Telecommunications Board Commission on Physical Sciences,*

Mathematics, and Applications. National Research Council, National Academy Press: US.

OECD (1992). *Guidelines for the Security of Information Systems*. OECD/GD (92) 190. Paris.

Olson, I. M., Abrams, M. D. (1995). Information security policy. In: Abrams, M. D., Jajodia, S., Podell, H. J., eds. *Information Security—an Integrated Collection of Essays*. Los Alamitos, CA: IEEE Computer Society Press.

Overly, M. R. (1999). *E-Policy: How to Develop Computer, E-Mail, and Internet Guidelines to Protect Your Company and Its Assets*. New York: AMACOM.

Pethia, R., Paller, A., Spafford, G. (2000). *Consensus Roadmap for Defeating Distributed Denial of Service Attacks*. USA: Global Institute Analysis Center, SANS Institute. http://www.sans.org/dosstep/roadmap.php (last accessed March 15 2003).

Rannenberg, K. (1994). Recent development in information technology security evaluation—the need for evaluation criteria for multilateral security. In: Sizer, R., Yngstrom, L., Kaspersen, H., Fischer-Hubner, S., eds. Proceeding of Security and Control of Information Technology in Society. IFIP Transactions A43. Elsevier Science B.V. (North-Holland).

Saunders Thomas, D., Forcht, K. A., Counts, P. (1998). Legal considerations of Internet use—issues to be addressed. *Internet Research* 8(1).

Stanley, A. K. (1997). Information security-challenges for the next millennium. In: Yngstrom, Y., Carlsen, J., eds. *Information Security in Research and Business*. Proceedings of the 13th International Conference on Information Security (SEC #97, IFIP. Denmark: Chapman & Hall.

Yngstrom, L. (1995). A holistic approach to IT security. In: Eloff, J. H. P., Von Solms, H. S., eds. *Information Security—the Next Decade*, IFIP/Sec '95. Proceedings of the IFIP TC11 11th International Conference on Information Security. Chapman and Hall.

2

Managing Software Security Risks: Buffer Overflows and Beyond

GARY McGRAW
Cigital, Dulles, VA, USA

1. SOFTWARE SECURITY

Vulnerable software is the biggest problem in computer security today; yet most of the money spent managing computer security risk is spent on firewalls, cryptography, and antivirus protection. Typical organizations invest in security by buying and maintaining a firewall, but go on to let anybody access remotely exploitable Internet-enabled applications through the firewall. A similar overemphasis on cryptography leads many organizations to ignore other critical aspects of software security.

Attackers exploit software. Making progress on the computer security problem depends on making software behave.

Current approaches, based on fixing things only after they have been exploited in fielded systems, address only symptoms, ignoring root causes. Proactively treating software and software development as a risk management problem is a superior approach.

2. UNDERSTANDING THE PROBLEM

The software security problem is growing. Three major trends appear to be driving the problem. The current accepted standard of fixing broken software only after it has been compromised (often called "penetrate and patch") is insufficient to control the problem.

Security holes in software are common. The frequency and magnitude of CERT Alerts and postings to the security mailing list bugtraq show how fast the problem is growing (with around 20 new vulnerabilities made public each week). Even "tried and true" software may not be as safe as one might think; many vulnerabilities that have been discovered in software existed for months, years, and even decades before discovery.

Most modern computing systems are susceptible to software security problems, so why is software security a bigger problem now than in the past? Three major trends have changed the classic risk environment that software exists in.

Networks are everywhere: The growing connectivity of computers through the Internet has increased both the number of attack vectors, and the ease with which an attack can be made. This puts software at greater risk. People, businesses, and governments are increasingly dependent upon network-enabled communication provided by information systems. Unfortunately, as critical systems are connected to the Internet, they become vulnerable to software-based attacks from distant sources. An attacker no longer needs physical access to a system to exploit software.

Systems are easily extensible: An extensible host accepts updates or extensions, sometimes referred to as mobile code, so that the functionality of the system can be evolved in an incremental fashion (McGraw and Felten, 1999). Sun

Microsystem's Java and Microsoft's NET framework aim to make this even more common. Today's operating systems support extensibility through dynamically loadable device drivers and modules. Today's applications, such as word-processors, e-mail clients, spreadsheets, and Web-browsers support extensibility through scripting, controls, components, and applets. Unfortunately, the very nature of extensible systems makes preventing software vulnerabilities from slipping in as an unwanted extension a challenge.

System complexity is rising: A desktop system running Windows XP and associated applications depends upon the proper functioning of the kernel as well as the applications to ensure that vulnerabilities cannot compromise the system. However, XP itself consists of at least forty million of lines of code, and end user applications are becoming equally, if not more, complex. When systems become this large, bugs cannot be avoided. Figure 1 shows how the complexity of Windows (measured in lines of code) has grown over the years.

The complexity problem is exacerbated by the use of unsafe programming languages (e.g., C or C++) that do not

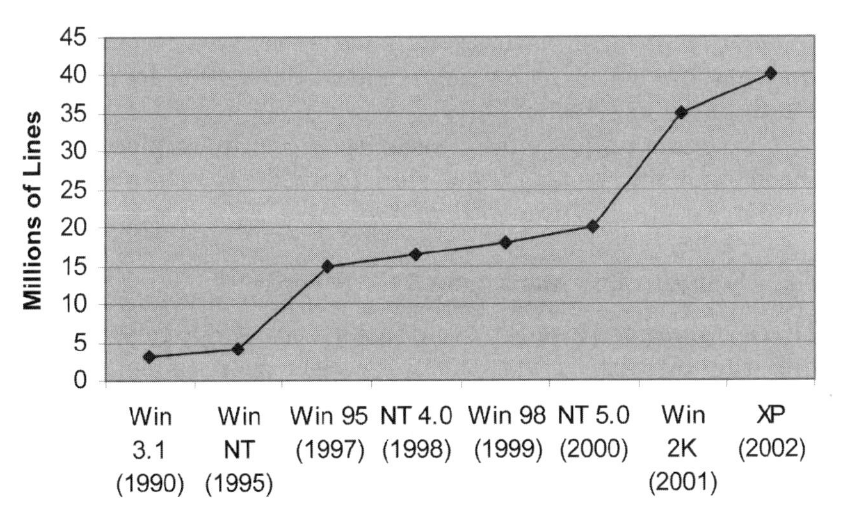

Figure 1 Growth of the Microsoft Operating System code base.

protect against simple kinds of attacks, such as buffer overflows. In theory, we could analyze and prove that a small program was free of problems, but this task is impossible for even the simplest desktop systems today, much less the enterprise-wide systems used by businesses or governments.

2.1. Security Software != Software Security

Many software vendors fail to understand that security is not an add-on feature. They continue to design and create products with little attention paid to security. When they do address security, it is often by adding security features, not understanding that software security is not security software.

Vendors start to worry about security only after their product has been publicly (and often spectacularly) broken by someone. Then they rush out a patch instead of coming to the realization that designing security in from the start might be a better idea. The unfortunately pervasive "penetrate and patch" approach to security is quite poor, suffering from problems caused by desperately trying to come up with a fix to a problem that is being actively exploited by attackers. In simple economic terms, finding and removing bugs in a software system before its release is orders of magnitude cheaper than trying to fix systems after release. Realize that each patch released by a vendor is an "attack map" for exploiting unpatched systems.

Designing a system for security, carefully implementing the system, and testing the system extensively before release, presents a much better alternative.

2.2. Software Risk Management for Security

There is no substitute for working software security as deeply into the software development process as possible, taking advantage of the engineering lessons software practitioners have learned over the years. Software engineering provides many useful tools that good software security can leverage. The key to building secure software is treating software security as risk management and applying the tools in a manner that is consistent with the purpose of the software itself (Viega and McGraw, 2001).

The aphorism, "keep your friends close and your enemies closer", applies quite aptly to software security. Risk identification and risk assessment are thus critical. The key to an effective risk assessment is expert knowledge of security. Being able to recognize situations where common attacks can be applied is half the battle. Software security risks come in two main flavors: architectural problems and implementation errors. Most software security material orbits exclusively around implementation errors. These issues are important, but focusing solely on the implementation level will not solve the software security problem.

Building secure software is like building a house. Correct use of security-critical system calls (such as string manipulation calls in C and C++) is like using solid bricks as opposed to using bricks made of sawdust. The kinds of bricks you use are important to the integrity of your house, but even more important (if you want to keep bad things out) is having four walls and a roof in the design. The same thing goes for software: what system calls you use and how you use them is important, but overall design properties often count for more.

2.3. Implementation Risks

Though simply ferreting known problems out of software implementations is not a complete solution, implementation risks are an important class of problems that deserve attention. Seven common implementation level problems are as follows:

1. Buffer overflows: Buffer overflows have been causing serious security problems for decades (Wagner et al., 2000). Buffer overflows accounted for over 50% of all major security bugs resulting in CERT/CC advisories in 1999. The root cause behind buffer overflow problems is that C is inherently unsafe (as is C++). There are no bounds checks on array and pointer references and there are many unsafe string operations in the standard C library. For these reasons, it is

imperative that C and C++ programmers writing security critical code learn about the buffer overflow problem.

2. Race conditions: Race conditions are possible only in environments where there are multiple threads or processes occurring at once that may potentially interact (or some other form of asynchronous processing). Race conditions are an insidious problem, because a program that seems to work fine may still harbor them (Bishop and Dilger, 1996). They are very hard to detect, especially if you are not looking for them. They are often difficult to fix, even when you are aware of their existence. In a world where multithreading, multiprocessing, and distributed computing are becoming more and more prevalent, race conditions will continue to be a problem.

3. Access control problems: Once users have successfully authenticated to a system, the system needs to determine what resources each user should be able to access. There are many different access control models for answering that question. Some of the most complicated are used in distributed computing architectures and mobile code systems, such as the CORBA and Java's EJB models. Mis-use of complex access control systems is a common source of software security problems.

4. Randomness problems: Random numbers are important in security for generating cryptographic keys and many other things. Assuming that C's rand() and similar functions produce unpredictable result is natural but unfortunately flawed. A call to rand() is really a call to a traditional "pseudo-random" number generator (PRNG) that is quite predictable.

5. Misuse of cryptography: One sweeping recommendation applies to every use of cryptography: *Never "roll your own" cryptography*! The next most commonly encountered crypto mistakes include failing to apply cryptography when it is really called for, and incorrect application of cryptography even when the need has been properly identified.

6. Input validation mistakes: Software architects have a tendency to make poor assumptions about who and what they can trust. Trust is not something that should be extended lightly. Sound security practice dictates the assumption that

everything is untrusted by default. Trust should only be extended out of necessity.

7. Password problems: Every computer user knows what a password is. Like many security technologies, the concept of a password is simple and elegant, but getting everything exactly right is much harder than it first appears. Two areas of particular vulnerability include password storage and user authentication with passwords.

Note that there are many other implementation level software security problems. The seven outlined here are most commonly encountered.

Static analysis tools like ITS4 (http://www.cigital.com/its4) and SourceScope (a parser-based approach) can be used to find many implementation problems in source code (Viega et al., 2000). Rule sets for C, C++, and (to a lesser extent) Java are available and in common use today. Research to determine how well implementation level vulnerabilities can be automatically fixed by syntactic transformation (using Aspect Oriented Programming) is ongoing. Runtime solutions to implementation problems are also available, but are not as effective.

2.4. Architecture Guidelines

Architectural analysis is more important to building secure software than implementation analysis is. It is also much harder and much less well understood. Following a simple set of general principles for developing secure software systems can help. Proper software security at an architectural level requires careful integration of security into a mature software engineering process.

The biggest open research issue in software security is that there is currently no good standard language of discourse for software design. Lacking the ability to specify an application formally, tools and technologies for automated analysis of software security at the architectural level lag significantly behind implementation tools. Until the research community makes more progress on this issue, architectural risk analysis will remain a high expertise practice.

Building Secure Software presents a list of 10 design level guidelines, reproduced here without commentary. Following these principles should help developers and architects who are not necessarily versed in security avoid a number of common security problems.

1. Secure the weakest link.
2. Practice defense in depth.
3. Fail securely.
4. Follow the principle of least privilege.
5. Compartmentalize.
6. Keep it simple.
7. Promote privacy.
8. Remember that hiding secrets is hard.
9. Be reluctant to trust.
10. Use your community resources.

Some caveats are in order. As with any complex set of principles, there are often subtle tradeoffs involved in their use. There is no substitute for experience. A mature software risk management approach provides the sort of data required to apply the principles intelligently.

3. A CALL TO ARMS

The root of most security problems is software that fails in unexpected ways. Though software security as a field has much maturing to do, it has much to offer to those practitioners interested in striking at the heart of security problems.

Good software security practices can help ensure that software behaves properly. Safety-critical and high assurance system designers have always taken great pains to analyze and track software behavior. Security-critical system designers must follow suit.

Software practitioners are only now becoming aware of software security as an issue. Plenty of work remains to be done in software security. The most pressing current need involves understanding architectural-level risks and flaws.

REFERENCES

Bishop, M., Dilger, M. (1996). Checking for race conditions in file access. *Computing Systems* 9(2):131–152.

McGraw, G., Felten, E. (1999). *Securing Java: Getting Down to Business With Mobile Code*. New York, NY: John Wiley & Sons. See http://www.securingjava.com/.

McGraw, G., Morrisett, G. (2000). Attacking malicious code: a report to the Infosec Research Council. *IEEE Software* 17(5).

Viega, J., Bloch, J. T., Kohno, T., McGraw, G. (2000). ITS4: a static vulnerability scanner for C and C++ code. In: Proceedings of Annual Computer Security Applications Conference. New Orleans, LA, December.

Viega, J., McGraw, G. (2001). *Building Secure Software*. New York: Addison-Wesley. See http://www.buildingsecuresoftware.com/.

Wagner, D., Foster, J., Brewer, E., Aiken, A. (2000). A first step towards automated detection of buffer over-run vulnerabilities. In: Proceedings of the Year 2000 Network and Distributed System Security Symposium (NDSS). San Diego, CA.

3

Experiments in Wireless Internet Security*

VIPUL GUPTA
Sun Microsystems Laboratories,
Mountain View, CA, USA

SUMIT GUPTA
Sun Microsystems Laboratories,
Mountain View, CA, USA

ABSTRACT

Internet enabled wireless devices continue to proliferate and are expected to surpass traditional Internet clients in the near future. This has opened up exciting new opportunities in the mobile e-commerce market. However, data security and privacy remain major concerns in the current generation of "wireless web" offerings. All such offerings today use a security architecture that lacks end-to-end security. This

*This article originally appeared in Proceedings of the IEEE Wireless Communications and Networking Conference, 2002.

unfortunate choice is driven by perceived inadequacies of standard Internet security protocols like Secure Sockets Layer (SSL) on less capable CPUs and low bandwidth wireless links. This paper presents our experiences in implementing and using standard security mechanisms and protocols on small wireless devices. Our results show that SSL is a practical solution for ensuring end-to-end security of wireless Internet, transactions even within today's technological constraints.

1. INTRODUCTION

The past few years have seen an explosive growth in the popularity of small, handheld devices (mobile phones, PDAs, pagers), that are wirelessly connected to the Internet. These devices, which are predicted to soon outnumber traditional Internet hosts like PCs and workstations, hold the promise of ubiquitous ("anytime, anywhere") access to a wide array of interesting services. However, limitations of these battery-driven devices like small volatile and non-volatile memory, minimal computational capability, and small screen sizes, make the task of creating secure, useful applications for these devices especially challenging. It is easy to imagine a world in which people rely on connected handheld devices not only to store their personal data, check news and weather reports, but also for more security sensitive applications like on-line banking, stock trading, and shopping—all while being mobile. Such transactions invariably require the exchange of private information like passwords, PINs, and credit card numbers and ensuring the secure transport of this information through the network becomes an important concern.*

---

*While protecting data on the device is also important, mechanisms doing so are already available and not discussed further in this paper.

On the wired Internet, Secure Sockets Layer (SSL) (Frier et al.) is the most widely deployed and used security protocol.† It often takes years of widespread public review and multiple iterations (Needham and Schroeder, 1978, 1990) to discover and correct subtle but fatal errors in the design and/or implementation of a security protocol. Over the years, the SSL protocol and its implementations have been subjected to careful scrutiny by security experts (Wagner and Schneier, 1996). No wonder then that today SSL is trusted to secure sensitive applications ranging from web banking to online trading to all of e-commerce. The addition of SSL capabilities to mobile devices would bring the same level of security to the wireless world. Unfortunately, none of the popular wireless data services today offer SSL on a handheld device. Driven by perceived inadequacies of SSL in a resource constrained environment, architects of both WAP and Palm.net chose to use a proxy-based architecture, which is depicted in Figure 1(a). In this approach, a different security protocol (incompatible with SSL) is used between the mobile client and the proxy/gateway, e.g., WAP uses WTLS, (WAP Forum), and Palm.Net uses a proprietary protocol on the wireless link. The proxy then decrypts encrypted data sent by a WAP phone using WTLS and re-encrypts it using SSL before forwarding it to the eventual destination server. The reverse process is used

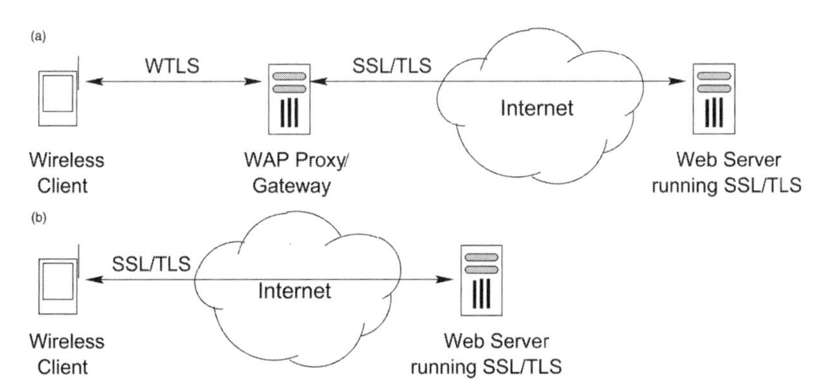

Figure 1 (a) Proxy-based architecture (top), and (b) end-to-end architecture (bottom).

for traffic flowing in the opposite direction. Such a proxy-based architecture has some serious drawbacks. The proxy is not only a potential performance bottleneck, but also represents a "man-in-the-middle" which is privy to all "secure" communications.

Even though the data are encrypted on both the wireless and wired hops, anybody with access to the proxy itself could see the data in clear text. Sometimes the situation is even worse and no encryption is used on the wireless side. This lack of end-to-end security is a serious deterrent for any organization thinking of extending a security sensitive Internet-based service to wireless users. Banks and brokerage houses are uncomfortable with the notion that the security of their customers' wireless transactions depends on the integrity of the proxy under the control of an untrusted third party. Moreover, the proxy is often pre-programmed into the clients' devices, which may raise legal issues (Bradner, 2000). For instance, Palm. Net users have to go through the proxy owned and operated by Palm. In contrast, the use of SSL between desktop PCs or workstations and Internet servers offers true end-to-end security (Figure 1(b)). This holds true even when a HTTPS proxy is used to traverse firewalls. Unlike the WAP or Palm.net proxy, a HTTPS proxy only acts as a simple TCP relay shuttling encrypted bytes from one side to the other without decryption/re-encryption. We felt that the claims about the unsuitability of SSL for mobile devices had not been adequately substantiated (Miranzadeh). This prompted our experiments in evaluating SSL (considered too "big" by some) for small devices. We sought answers to some key questions: is it possible to develop a usable implementation of SSL for a mobile device and thereby provide end-to-end security? How would near-term technology trends impact the conclusions of our investigation? The rest of this paper describes our experiments in greater detail. Section 2 provides a brief overview of the SSL protocol. Section 3 discusses our implementation of an SSL client, called KSSL, on a Palm PDA and evaluates its performance. Section 4 describes an application we have developed for secure, mobile access to enterprise resources based on KSSL.

Section 5 talks about mobile technology trends relevant to application and protocol developers. Finally, we offer our conclusions in Section 6.

2. SECURE SOCKETS LAYER (SSL)

2.1. Overview

SSL offers encryption, source authentication, and integrity protection of application data over insecure, public networks. Figure 2 shows the layered nature of this protocol.

The Record layer, which sits above a reliable transport service like TCP, provides bulk encryption and authentication using symmetric key algorithms. The keys for these algorithms are derived from a master secret established by the Handshake protocol between the SSL client and server using public-key algorithms. SSL is very flexible and can accommodate a variety of algorithms for key agreement, encryption, and hashing. To guard against adverse interactions (from a security perspective) between arbitrary combinations of these algorithms, the standard specification explicitly lists combinations of these algorithms, called cipher-suites, with well-understood security properties. The Handshake protocol is the most complex part of SSL with many possible variations (Figure 3). In the following section, we focus on its most popular form, which uses RSA key exchange and does not

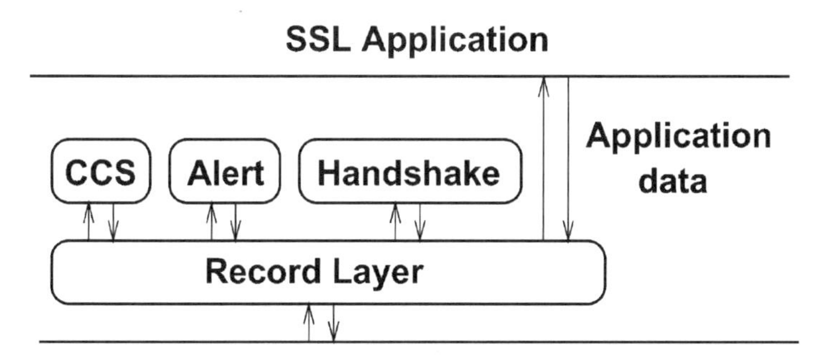

Figure 2 SSL architecture.

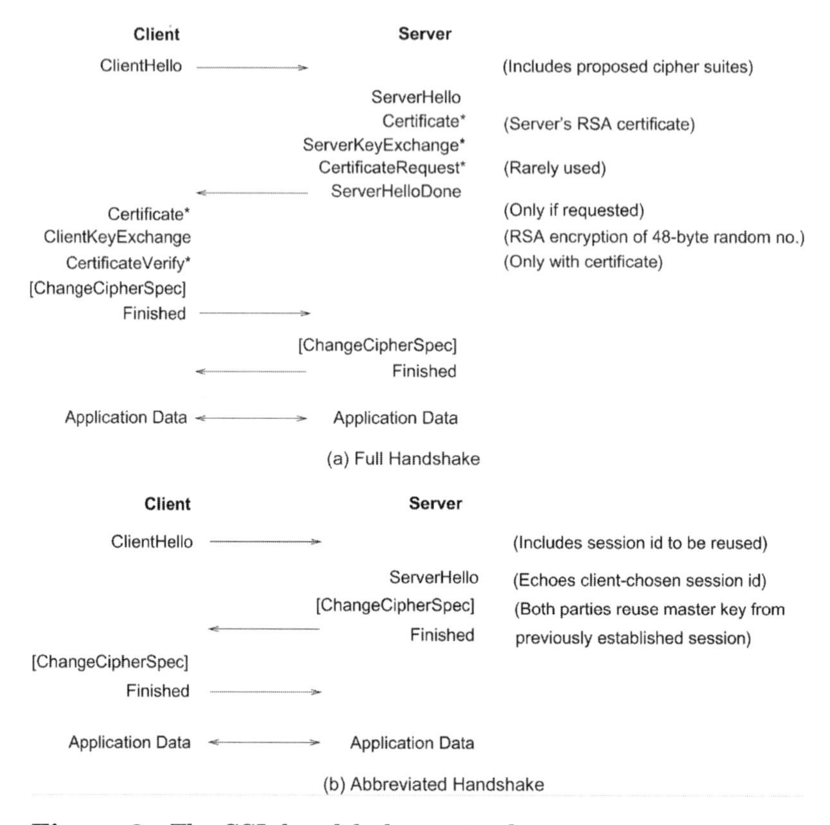

Figure 3 The SSL handshake protocol.

involve client-side authentication. While SSL allows both client- and server-side authentication, only the server is typically authenticated due to the difficulty of managing client-side certificates. Client authentication, in such cases, happens at the application layer, e.g., through the use of passwords sent over an SSL-protected channel.

2.1.1. Full SSL Handshake

When an SSL client encounters a new server for the first time, it engages in the full handshake shown in Figure 3(a). The client and server exchange random nonces (used for replay protection) and negotiate a mutually acceptable cipher suite in the first two messages. The server then sends its RSA public-key inside an X.509 certificate. The client verifies the public-key, generates a 48-byte random number (the pre-master secret) and sends it encrypted with the server's public-key. The server uses its RSA private-key to decrypt the pre-master secret. Both end-points use the pre-master secret to create a master secret, which, along with previously exchanged nonces, is used to derive the cipher keys, initialization vectors, and MAC (Message Authentication Code) keys for the Record layer.

2.1.2. Abbreviated SSL Handshake

A client can propose to reuse the master key derived in a previous session by including that session's (non-zero) ID in the first message.[‡] The server indicates its acceptance by echoing that ID in the second message. This results in an abbreviated handshake without certificates or public-key cryptographic operations so fewer (and shorter) messages are exchanged (see Figure 3(b)). An abbreviated handshake is significantly faster than a full handshake.

2.2. SSL on Small Devices: Common Perception vs. Informed Analysis

SSL is commonly perceived as being too heavy weight for the comparatively weak CPUs in mobile devices and their low bandwidth, high latency wireless networks. The need for RSA operations in the handshake, the verbosity of X.509 encoding, the chattiness (multiple round trips) of the handshake, and the large size of existing SSL implementations are all sources of concern.

However, we are not aware of any empirical studies evaluating SSL for small devices and careful analysis of the protocol's most common usage reveals some interesting insights:

- Some constraints ease others. If the network is slow, the CPU does not need to be very fast to perform bulk encryption and authentication at network speeds.

[‡]The server can also force a full-handshake by returning a new session ID.

- A typical SSL client only needs to perform RSA public-key, rather than private-key, operations for signature verification and encryption. Their small exponents (typically no more than 65537, a 17-bit value) make public-key operations much faster than private-key operations. It is worth pointing out that the performance of RSA public-key operations is comparable to that of equivalent Elliptic Curve Cryptography (Koblitz, 1994) operations (Boneh and Daswani, 1999).
- There are several opportunities to amortize the cost of expensive operations across multiple user transactions. Most often, a client communicates with the same server multiple times over a short period of time, e.g. such interaction is typical of a portal environment. In this scenario, SSL's session reuse feature greatly reduces the need for a full handshake. Although SSL can be used to secure any connection oriented application protocol (SMTP, NNTP, IMAP), it is used most often for securing HTTP. The HTTP 1.1 specification encourages multiple HTTP transactions to reuse the same TCP connection. Since an SSL handshake is only needed immediately after TCP connection set up, this "persistent HTTP" feature further decreases the frequency of SSL handshakes.

3. KSSL AND KSECURITY

"KiloByte" SSL (KSSL) is a small footprint, SSL client for the Mobile Information Device Profile (MIDP) of Java 2 Micro-Edition (J2ME™) (see MIDP). Its overall architecture and relationship to the base J2ME platform is depicted in Figure 4. The KSecurity package provides fundamental cryptographic functions such as random number generation, encryption, and hashing that are missing from base J2ME. It reuses the Java Card™ API which opens up the possibility of using a JavaCard as a hardware crypto accelerator with minimal changes to the KSSL code. Some of the compute

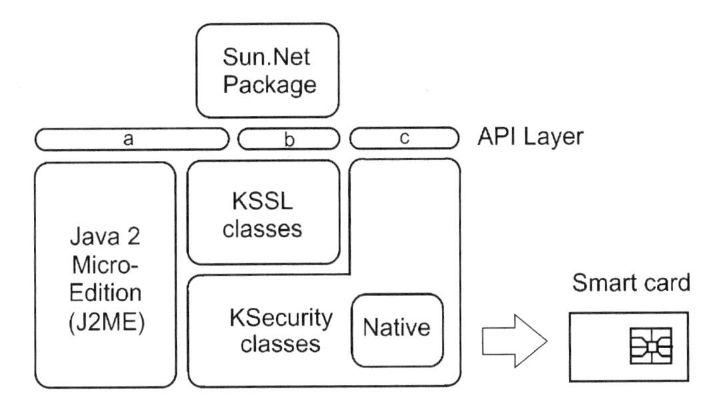

Figure 4 KSSL implementation architecture.

intensive operations (such as modular exponentiation) are implemented as native methods in C. An SSL client also needs to process X.509 certificates and maintain a list of trusted certificate authorities. Since the Java Card APIs do not deal with Certificates or Keystones, we modeled these classes as subsets of their Java 2 Standard Edition (J2SE™) counterparts. The SSL protocol (box labeled KSSL) is written purely in Java™ and implements the client-side of SSLv3.0, the most popular version of SSL. It offers only two cipher suites RSA_RC4_128_MD5 and RSA_RC4_40_MD5 since they are fast and almost universally implemented by SSL servers. Client-side authentication is not implemented. KSSL interoperates successfully with SSL servers from iPlanet™, Microsoft, Sun™, and Apache (using OpenSSL).

3.1 Performance

3.1.1. Static Memory Requirements

For PalmOS, the addition of KSSL and KSecurity classes increases the size of the base J2ME implementation by about 90 KB. This additional memory is reasonable compared to the size of base J2ME which is typically a few hundred KB. It is possible to reduce the combined size of KSSL and Ksecurity packages to as little as 70 KB if one is willing to sacrifice the clean interface between them (applications will only be able to access security services through SSL).

3.1.2. Bulk Encryption and Authentication

Table 1 shows the timing results of some cryptographic operations on the Palm.

We found that the bulk encryption and authentication algorithms are adequately fast even on the Palm's CPU. On a 20 Mhz chip (found in Palm Vx, Palm IIIc, etc.), RC4, MD5, and SHA all run at over 100 Kbits/sec, since each SSL record requires both an MAC computation and encryption, the effective speed of bulk transfer protected by RC4/MD5 is well over 50 Kbits/sec, far more than the 9.6 Kbits/sec bandwidth offered by our Omnisky CDPD network service.

SSL Handshake Latency

A typical handshake requires two RSA public-key operations: one for certificate verification and another for premaster secret encryption. As shown in Table 1, our implementation of RSA takes 0.8–1.5 sec on a 20 MHz Palm CPU, depending on the key size (768 bits or 1024 bits). Other factors such as network delays and the time to parse X.509

Table 1 Performance of KSSL Cryptographic Primitives on PDA

	Palm Vx (20 MHz)	Visor (33 MHz)
RSA (1024-bit)		
Verify[a]	1433 ms	806 ms
Sign	80.91 sec	45.1 sec
RSA (768-bit)		
Verify[a]	886 ms	496 ms
Sign	36.22 sec	20.19 sec
MDS		
1024 bytes	292 Kbits/sec	512 Kbits/sec
4096 bytes	364 Kbits/sec	655 Kbits/sec
SHA-1		
1024 bytes	124 Kbits/sec	227 Kbits/sec
4096 bytes	140 Kbits/sec	256 Kbits/sec
RC4		
1024 bytes	117 Kbits/sec	215 Kbits/sec
4096 bytes	190 Kbits/sec	351 Kbits/sec

[a] With a public-key exponent of 65537.

certificates also impact the handshake latency. In our experiments with Palm MIDP, we found that a full handshake can take approximately 10 sec. In most scenarios, a client communicates with the same SSL server repeatedly over a small duration. In such cases, the handshake overhead can be reduced dramatically. For example, simply caching the server certificate (indexed by an MD5 hash) eliminates the overhead of certificate parsing and verification. This brings down the latency of a subsequent full handshake to around 7–8 sec. An abbreviated handshake only takes around 2 sec. Finally, by using persistent HTTP, one could make the amortized cost of an SSL handshake arbitrarily close to 0.

4. SECURE MOBILE ENTERPRISE ACCESS

To evaluate usability of the KSecurity and KSSL packages for "real-world" applications, we have developed a J2ME midlet suite (an MIDP application is called a midlet) that enables Sun employees to securely access enterprise services like corporate email, calendar, and directory from a PalmVx connected to the Internet via Omnisky's wireless CDPD modem. The application size is about 55 KB and it runs in 64 KB of application heap.

The overall architecture shown in Figure 5 reuses Sun's existing SSL-based virtual private network (VPN) for remote employees, which is based on the iPlanet Portal Server. All communication between the VPN gateway and the mobile device is protected end-to-end by SSL. Remote users must authenticate themselves through a challenge–response mechanism (using tokencards) over an SSL-secured channel before they can access the Intranet. Authenticated users are allowed to communicate with specially designed servlets running on a web server behind the firewall. The main purpose of these servlets is to format data obtained from mail, calendar or LDAP servers for consumption by the remote application. The servlets are capable of maintaining session-specific state. This allows for features such as "chunking" where a long email message is sent to the user in smaller

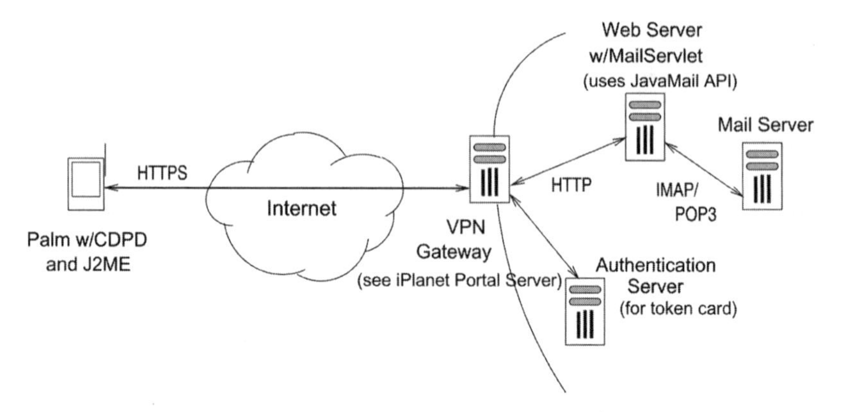

Figure 5 Secure email client architecture.

parts for bandwidth efficiency (this avoids sending any long messages in which the user has no interest). By making effective use of certificate caching and SSL session reuse, we are able to reduce the response time of each user transaction to about 8 sec which is comparable to the response time of accessing public web pages in the clear. For now, the persistent HTTP feature has not been turned on at the VPN gateway. Exploiting this feature would reduce the response time to around 5 sec.

5. TECHNOLOGY TRENDS

Since starting this project about a year ago, we have seen several examples of technology's relentless march towards smaller, faster, and more capable devices. Newer Palm PDAs like the PalmVx and PalmIIIc use 20 MHz processors and the Handspring Visor Platinum (another PalmOS device) features a 33 MHz processor, both considerable improvements over the earlier 16 MHz CPUs. All of them offer 8 MB of memory. The Compaq iPaQ pocket PC, in comparison, carries a 200 MHz StrongARM processor and 16–32 MB of memory. The CPU enhancements have a direct impact on the speed of SSL's cryptographic operations. Significant performance gains are also obtainable by using hardware accelerators in the form of tiny smart cards (and related devices like the iButton). The Schlumberger Cyberflex smart card, for instance, can perform 1024-bit RSA operations (both public- and private-key) in under one second! Similarly, improvements can also be seen in the speed of wireless networks. Metricom's Ricochet service now offers wireless data speeds of 128 Kbits/sec in several US cities and 3G networks hold the promise of even faster communication in the next year or two. These improvements help reduce the network-related latency of an SSL handshake. Even smart compilation techniques, which had so far been available only on more capable PCs and workstations, are now available on small devices and can boost the performance of J2ME applications by as much as a factor of 5. These developments should alleviate any remaining concerns about SSL's suitability for wireless devices. They also highlight an interesting phenomenon. In the time it takes to develop and deploy new (incompatible) protocols, technology constraints can change enough to raise serious questions about their longterm relevance. The following quote by Wayne Gretzsky (Ice Hockey Legend) captures this sentiment rather well: "Don't skate to the puck; skate to where it's going". In light of this view, it is reassuring to see the WAP Forum embracing standard IETF and W3C protocols for its next (WAP2.0) specification.

6. CONCLUSIONS AND FUTURE WORK

Our experiments show that SSL is a viable technology even for today's mobile devices and wireless networks. By carefully selecting and implementing a subset of the protocol's many features, it is possible to ensure acceptable performance and compatibility with a large installed base of secure web servers while maintaining a small memory footprint.

Our implementation brings mainstream security mechanisms, trusted on the wired Internet, to wireless devices for the first time. The use of standard SSL ensures end-to-end security, an important feature missing from current wireless architectures. Sun's MIDP reference implementation incorporating KSSL can be downloaded as part of the J2ME Wireless Toolkit from java.sun.com/products/midp.

In our ongoing effort to further enhance cryptographic performance on small devices, we plan to explore the use of smartcards as hardware accelerators and Elliptic Curve Cryptography in our implementations.

ACKNOWLEDGMENTS

We based portions of our native code on Ian Goldberg's PalmOS port of the crypto algorithms from SSLeay. We would also like to thank Nagendra Modadugu from Stanford University, and Sheueling Chang from Sun Microsystems for their help with optimizing RSA. Sun, Sun Microsystems, Java, Java Card, J2SE, J2ME, and iPlanet are trademarks or registered trademarks of Sun Microsystems, Inc., in the United States and other countries.

REFERENCES

Boneh, D., Daswani, N. (1999). Experiments with electronic commerce on the PalmPilot. *Financial Cryptography '99*. Lecture Notes in Computer Science. Vol. 1648. pp. 1–16.

Bradner, S. The problems with closed gardens. Network World, June 12, 2000, at http://www.nwfusion.com/columnists/2000/0612bradner.html.

Frier, A., Karlton, P., Kocher, P. (1996). The SSL Protocol Version 3.0. See http://wp.netscape.com/eng/ssl3/draft.302.txt.

Koblitz, N. (1994). *A Course in Number Theory and Cryptography*. 2nd Ed., Springer-Verlag.

Miranzadeh, T. Understanding security on the wireless internet. See http://www.tdap.co.uk/uk/archive/billing/bill (phonecom_9912).html.

Mobile Information Device Profile (MIDP) of Java 2 Micro-Edition (J2ME). See http://java.sun.com/products/midp.

Needham, R. M., Schroeder, M. D. (1978). Using encryption for authentication in large networks of computers. *Communications of the ACM* 21(12):993–999.

Needham, R. M., Schroeder, M. D. (1990). Authentication revisited. *Operating System Review* 21(1):p. 7.

Wagner, D., Schneier, B. (1996). Analysis of the SSL 3.0 protocol. Proceedings of the USENIX Workshop on Electronic Commerce. Available from http://www.cs.berkeley.edu/˜daw/papers/.

WAP Forum. Wireless transport layer security specification. See http://www.wapforum.org/what/technical.htm.

4

The Neverending Saga of Internet Security: Why? How? and What to Do Next?*

EDDIE RABINOVITCH
ECI Technology, East Rutherford, NJ, USA

Just in January 2001, our installment of "Your Internet Connection" (http://www.comsoc.org/ci/public/2001/jan/ciint.html) was dedicated to security standardization. So why did we decide to have yet another discussion on Internet security shortly after? Simply, because the issue of security becomes more important and critical, recently gaining additional exposure not just to the technical audience but to the general public as well. Let us face it: unless caught and forced

*Copied with permission of the author and *IEEE Communications Magazine*.

to do so, companies would very rarely disclose the fact that their security was compromised. With this in mind, according to statistics published by Carnegie Mellons's CERT® Coordination Center a total of 82,094 security incidents were reported last year in 2002 (which represents almost 45% of the 182,463 total number of incidents reported during last 15 years (!) between 1988 and 2002). The number of vulnerabilities reported also grew in geometrical proportion in the last 3 years: from 1090 in 2000 to 2437 in 2001, to 4129 in 2002. Other disturbing statistical data come from a survey of 503 security practitioners by the FBI and the Computer Security Institute (CSI), published in April of last year cost of Internet-related crimes have a costly effect on government, academia, and business institutions. Following are some of the highlights of the CSI "2002 Computer Crime and Security Survey":

- 90% of respondents detected computer security breaches within the last year.
- 44% of the respondents were able to quantify their financial loses due to security breaches totaling close to $456 million.
- For the fifth year in a row, more respondents (74% this time) cited their Internet connection as a frequent point of attack vis-à-vis 33% who cited their internal systems as a point of attack.
- 40% detected system penetration from the outside.
- 40% detected denial of service attacks.
- 85% detected computer viruses.
- 38% suffered unauthorized access or misuse on their Web sites within the last 12 months.
- 70% of those attacked reported vandalism.
- 12% reported theft of transaction information.
- 6% reported financial fraud (vs. only 3% in 2000).

And since for obvious reasons most companies would not disclose at all or highly underestimate their security exposures, this figure is probably very conservative, to say the least, and the real financial losses caused by Internet security exposures are most definitely significantly higher.

Back in March of 1997, in one of the first installments of Your Internet Connection (http://www.comsoc.org/ci/public/1997/mar/internet_column.html), describing the reasons for Internet security concerns, I referred to "... the ever-growing number of sophisticated users who are 'surfing' the net, sometimes with a clear intention to break into someone's network either as a 'hobby' or for industrial espionage. In other words, some do it for money, some for pleasure..." Recent well (and not-so-well) publicized Internet security breaches certainly point to a seriously dangerous growth in both populations: the so-called script-kiddies as well as vicious Internet criminals. For example, in an unprecedented move in early March 2001, while still under investigation, the FBI released information about a series of economic extortion attacks that at that time had already hit more than 40 electronic banking and e-commerce sites. During these attacks more than 1 million credit cards were stolen, and criminals, using typical organized crime tactics, extorted money "to protect sites from 'other' hackers", promising to keep the site's credit card information confidential. The Center for Internet Security (CIS), jointly with SANS Institute, asked Steve Gibson of *Gibson Research* to develop a tool that would determine system's vulnerability and level of exposure to the aforementioned attack. This tool, PatchWork, can be obtained from *the CIS*.

Another series of Internet security exposures that recently gained publicity are domain name system (DNS) vulnerabilities. In some cases these were flawed network topology designs leaving multiple DNS servers on a single network segment exposed to a single point of failure. In other cases, it was the infamous Berkeley Internet Name Domain (BIND) flaw. However, despite all the publicity and attention to DNS security exposures, months later DNS still remained vulnerable in several Fortune 1000 companies in the United States and Europe, tested and surveyed by DNS security consultancy *Men and Mice*.

An additional Internet security challenge (a.k.a. problem) is the uncertainty of the identity of the entity on the other end of an Internet session. As someone said, "On the Internet no one knows you are a dog..."; no offense to dogs intended;-). Thus, verifying the identity of a Web site or an FTP server on the Internet is very important. The way to do this, of course, is by using certificates issued by trusted authorities. Unfortunately, as with anything dealing with security, this process is also not 100% foolproof. As we learned some time ago, *VeriSign*—one of the major certificate authorities—issued such certificates to someone masquerading as a Microsoft employee. The problem was discovered and the improper certificates were immediately revoked. However, it still represents a problem if the revocation list is not checked, which is not necessarily an automated process with the different software packages.

We should acknowledge the fact that Internet security vulnerabilities are never going to disappear. Yes, the technology is certainly getting more and more sophisticated and advanced; unfortunately, so do security perpetrators. Therefore, it is even more important to stay on top of the technology and implement the most sophisticated security protection tools. For example, for a very long time state-full Internet firewalls were considered performance log jams because of the need to inspect each and every packet. Recently, with the advent of so-called load-balanced firewall topologies, where a group of state-full Internet firewalls can be load-balanced between two layers of layer-7-aware load-balancing switches, firewall performance and scalability concerns were addressed. Please see the following network diagram for such an example: http://www.cervalis.com/facilities/network/networkdiagram.pdf. In addition to offering firewall performance and scalability relief, such a configuration also protects against several denial-of-service attacks, improving general security posture.

However, as was mentioned earlier, although a very significant one, technology is only one element of the security puzzle. Enforcement of sound security policies is the most important factor in protection. Assuming such policies were in place in all the different security incidents described above, someone somewhere did not follow or comply with a certain policy or rule. It is very important for any and all security policies to be accepted and enforced at all levels within an organization. Management must take steps to educate staff and make all employees aware of the need for security, and take all appropriate measures required to maintain it. In many security-conscious environments, the primary emphasis is placed on blocking external attacks. Too often, organizations fail to recognize the vulnerabilities that exist within their own environments, such as internal attacks from disgruntled employees and industrial espionage. Therefore, it is certainly a good idea to emphasize the importance and position of security to all employees in a positive manner.

Network security can be protected through a combination of high-availability network architecture and an integrated set of security access control and monitoring mechanisms. Recent incidents demonstrate the importance of monitoring security and filtering not only incoming traffic, but also the outbound traffic generated within the network. Defining a solid up-to-date information protection program, with associated access control policies and business recovery procedures, should be the first priority on the agenda of every networked organization. Specifically, a firm's information security posture—an assessment of the strength and effectiveness of the organizational infrastructure in support of technical security controls—has to be addressed through the following activities:

- auditing network monitoring and incident response;
- communications management;
- configurations for critical systems: firewalls/air-gaps, DNS, policy servers;
- configuration management practices;
- external access requirements and dependencies;
- physical security controls;
- risk management practices;
- security awareness and training for all organization levels;
- system maintenance;
- system operation procedures and documentation;
- application development and controls;
- authentication controls;
- network architecture and access controls;

- network services and operational coordination;
- security technical policies, practices, and documentation.

A properly designed network with comprehensive security policies aimed at high availability, recoverability, and data integrity institutes the necessary infrastructure to conduct any activities in a secure reliable fashion, regardless of whether the public Internet, extranets, or intranets are being utilized.

In the above referenced January 2001 installment of my Your Internet Connection column with the *IEEE Communications magazine*, we addressed some of the existing and developing security standards, i.e., an important technology for LAN security policing Extensible Authentication Protocol Over Ethernet—EAPOE—IEEE 802.1X

EAP, originally developed for Point-to-Point Protocol (PPP) and described in the RFC 2284 () was extended to Ethernet LANs by the IEEE. Among the advantages of authentication on the edges of enterprise network offered by the IEEE 802.1X, the standard cites:

- improved security;
- reduced complexity;
- enhanced scalability;
- superior availability;
- translational bridging;
- multi propagation;
- end stations manageability.

Although IEEE 802.1X is often being associated with security protection for wireless communication it is much more appropriate for wired protocols, where the physical cabling infrastructure is much more secure. IEEE 802.1X communications begins with an unauthenticated client device attempting to connect with an authenticator (i.e., IEEE 802.11 Access Point or wired-based switch). The Access Point/switch responds by enabling a port for passing only EAP packets from the client to an authentication server (e.g. RADIUS). Once the client device is positively authenticated, the Access Point/switch opens client's port for other types of traffic. IEEE 802.1X technology is currently being supported by a variety of networking gear vendors as well as many client software providers.

As IEEE 802.1X proved to be rather inadequate for securing wireless infrastructure the IEEE is finalizing a more appropriate security standard for wireless, such as the IEEE 802.11i. There are, however, means for securing wireless infrastructures today by using products specially designed for wireless security implementing other acceptable standards such as secure authentication, authorization, auditing and IPSec encryption. Utilizing such products would allow users building scalable and at the same time secure and highly available wireless infrastructures with seamless roaming for wireless users. Such interesting, innovative and at the same time well-proven technology can be found in family of security solutions for wireless networks by ReefEdge: http://www.reefedge.com.

As was discussed earlier, securing physical and networking layers is critical for building secure backbone. Nevertheless, all seven layers of the OSI are playing a very important role in security infrastructure. Some, justifiably so, take it even a layer higher, to the so-called layer 8, or the "political layer".

In conclusion, I would like to reference the very important activities of the CIS, whose membership is open not just to companies and organizations representing industry and academia, but also to individuals interested in and dedicated to Internet security. So if you have not done so yet, I very strongly urge you to contact the CIS and consider becoming an active member in this important organization. And, as usual, please feel free to contact me with any feedback, comments, and ideas at the editorial board of *IEEE Communications Magazine* or directly at eddie@reefedge.com.

5

Secure Compartmented Data Access over an Untrusted Network Using a COTS-Based Architecture

PAUL C. CLARK, MARION C. MEISSNER, and KAREN O. VANCE
SecureMethods, Inc., Vienna, VA, USA

ABSTRACT

In this paper, we present an approach to secure compartmented data access over an untrusted network using a secure network computing architecture. We describe the architecture and show how application-level firewalls and other commercial-off-the-shelf (COTS) products may be used to implement compartmentalized access to sensitive information and to provide access control over an untrusted network and in a variety of environments. Security-related issues and assumptions are discussed. We compare our architecture to other models of controlling access to sensitive data and draw conclusions about the requirements for high-security solutions for electronic business as well as DoD applications.

1. INTRODUCTION

Users of commercial applications as well as more traditional users of high-security applications, such as the DoD and intelligence community, increasingly require access to sensitive information over an untrusted wide area network (WAN) using high-assurance security mechanisms. Although many systems have been designed for this purpose (Choo, 1999; Dalton, 1999) , often they are highly specialized, nonscalable, expensive and cumbersome to use. Such systems can be difficult and costly to maintain and upgrade, since changes must be implemented and tested in a customized and highly specialized environment. Furthermore, operating system changes often require modifications and re-testing of system applications to ensure the entire system continues to conform to security requirements.

Historically, the intelligence community has been a forerunner in requiring and developing high-security network applications. By analyzing their groundbreaking work, other researchers and developers have arrived at requirements for the next generation of high-security applications for electronic business, healthcare data management, and a plethora of other commercial and noncommercial applications. These requirements include the following and form the starting point for the secure network computing architecture described in this paper.

- High-assurance security: Applications should provide high confidence through strong security services for confidentiality, data integrity, user-based authentication, and nonrepudiation. System assurance should be limited only by the underlying COTS platform and applications.

- Secure submission and retrieval: Data must be protected while allowing authorized users to access and update information.
- High usability: Uniform-user interfaces, such as those provided by e-mail and the World Wide Web, should be used to increase user acceptance and efficiency while reducing the training costs of high-security solutions.
- Scalability: A secure solution must be extensible to meet ever-increasing throughput and storage requirements.
- Commercial-off-the-shelf (COTS) products: Secure solutions should employ COTS-based hardware and software where possible to leverage vendor and industry development, testing, and validation and to minimize redevelopment and support efforts. COTS clients, for example, personal computers running Windows operating systems, are already prevalent but servers also benefit from the use of COTS products while maintaining the appropriate level of security assurance for each application.

Current firewall technologies and powerful, low-cost computing platforms lend themselves to a solution based on COTS hardware and software and standards-based cryptographic algorithms and mechanisms. In this paper we present such an approach to providing secure user-based access to protected data over an untrusted WAN. The deployed secure network architecture comprised of COTS components provides secure data access over trusted or untrusted networks and high-assurance access control in a highly usable, flexible, and scalable manner.

In our approach, transaction-oriented access to sensitive data is controlled on a per-user basis. Individual users are authenticated for each transaction using cryptographic techniques and authorizations are verified before data submissions or retrievals are permitted. The data being protected and securely accessed are compartmented in a manner that reflects the organization of data in the real world and/or in the application environment. This compartmentalized approach has been used successfully on standalone systems to restrict access to information on a "need to know" or project-specific basis. It also allows access to be controlled by specific resources and access rights, as well as on a per-project or per-compartment basis. Trusted servers on a protected network domain can store the data for different compartments and process requests for data accesses to that compartment whether requests are generated on the local network or remotely over the WAN.

This paper describes our approach and how the architecture satisfies the requirements of convenient, secure and tightly controlled access to sensitive data over a network. Then, we address security issues inherent in the architecture, compare it to related research work in the area, and, finally, discuss future directions for the architecture and its implementations.

2. SECURITY ARCHITECTURE

The architecture presented here uses a commercially available application-level firewall and other COTS machines running standard Unix or Windows NT/2000 operating systems to act as gateways between the users and the servers. The firewall mediates access between multiple network domains, with trusted servers and potentially several different levels of users in two or more separate domains. In the trusted server domain, a different server stores data and processes access requests for each of several compartments per transaction. These servers run unmodified COTS operating systems and software. By policy, only administrative user logins are permitted on this trusted server domain. The architecture therefore provides strong protection for sensitive data by physically separating servers containing compartmented information from other users and servers on the network. This configuration is illustrated in Figure 1.

As shown in Figure 1, users can use client workstations located in any of the user domains. They access

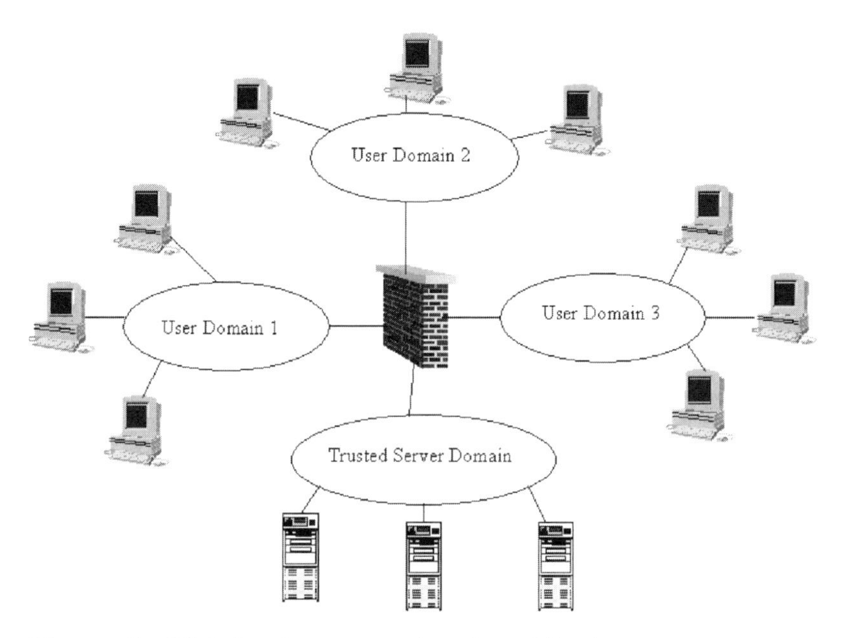

Figure 1 High-level secure network architecture.

compartmented information on the trusted servers via a user interface, such as a Web-based form. A protocol gateway, such as a Web server, located on an untrusted network domain, presents the user interface on the client workstation. Using a Web-based form, the user completes a secure data submission or retrieval request, and encrypts and digitally signs the request using cryptographic algorithms and optional hardware tokens, as appropriate for the application and compartment to be accessed. In addition to or instead of a digital signature, a user may utilize a biometric to provide authentication for the request. For higher usability, the enhancement process (encryption, signature, encoding, etc.) can occur automatically, for example, through a Web plugin. Depending on the application requirements, users optionally have to initiate or acknowledge enhancement of requests. The firewall relays the data submission or retrieval request to the trusted server on the appropriate network domain, via a cryptographic gateway.

Once the request reaches the trusted domain, the cryptographic gateway verifies the digital signature and/or biometric template and decrypts the transaction. The architecture supports Public Key Infrastructure (PKI) mechanisms and directories for managing and retrieving user certificates and cryptographic keys in a scalable manner, e.g., LDAP. After successful authentication, the cryptographic gateway checks the requested action and the originator against its access control list (ACL), and determines whether the user is authorized to perform the requested action. If the user's request is not authorized or the authentication fails, the cryptographic gateway will reject the request immediately. Authenticated requests from authorized users are delivered to the appropriate server to be processed. The server sends the result of the request to the cryptographic gateway, where it is encrypted for the originating user and digitally signed, and relayed back to the originator. Figure 2 shows the configuration necessary for an authenticated and authorized access request from User A on User Domain 2 to Compartment X.

Network accesses to the trusted server domain are strictly transaction-oriented and must be digitally signed

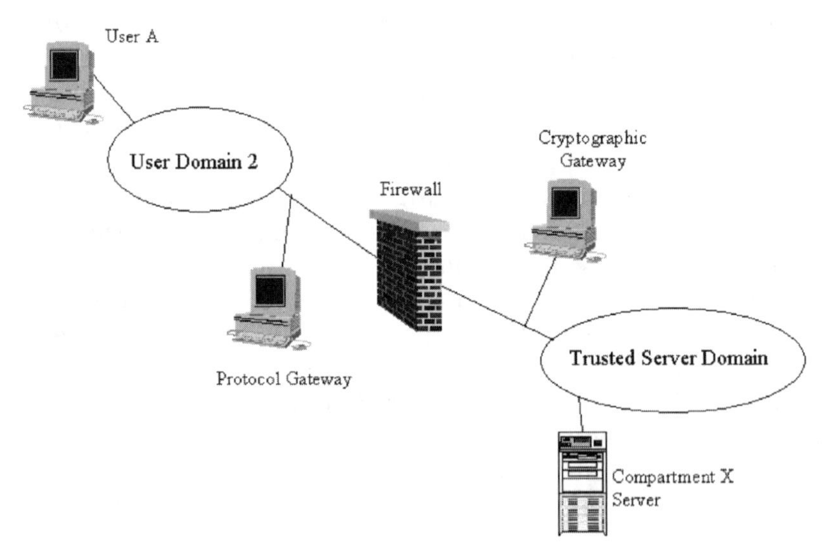

Figure 2 Configuration for a Sample Access Request.

and/or biometrically authenticated by the individual user making the request. The security policy for this architecture permits no direct network logins to trusted servers, except for administrative access. This policy is enforced by the firewall, which mediates access between the different network domains, allowing only transaction-oriented communications addressed to the cryptographic gateway to pass through to the trusted server domain. Administrative logins are protected using strong authentication. Thus, the firewall is used to control the perimeter of the trusted server domain.

As mentioned previously, authorizations are performed through the use of ACLs maintained on the cryptographic gateway located on the trusted server domain. In our approach, ACLs are human-readable files containing information on protected resources and the individual users that may access them. To be accessible from the WAN, each resource must be explicitly listed in one of the ACL files, along with authorized users. Resources are represented by Uniform Resource Locators (URLs), making them both specific and finely granular. The protocol used to access the resource, the hostname on which the resource is located, and the location and name of the resource can all be specified with this format. Since the firewall prevents direct connections to machines on the trusted domain, the ACLs are inaccessible to regular users and cannot be modified. Furthermore, the firewall routes all transactions to the cryptographic gateway to be verified and authorized against the ACL before being processed by a server. In this way, the architecture enforces access controls, equivalent to mandatory access control (MAC), for any compartmentalized data stored on the trusted servers.

As described above, the firewall, the cryptographic gateway, and the trusted servers on the server domain together enforce a unified security policy over the system. Thus, according to TCSEC and TNI criteria (National Computer Security Center, 1985, 1993), the firewall and the trusted network domain constitute a distributed trusted computing base (TCB) for the architecture—without the need for specialized operating systems. Together with an integrity protected, properly functioning client and the cryptographic mechanisms used to encrypt and digitally sign each transaction between the user and the cryptographic gateway, this TCB provides a trusted path between the user and the compartmentalized sensitive data storage at a level of assurance determined by the cryptographic algorithm, key length, and key management scheme employed.

The architecture described here can be scaled easily to meet the capacity and performance needs of different applications. As system loads increase, additional protocol gateways, cryptographic gateways and trusted compartmented data servers may be installed to provide increased capacity and fault tolerance through redundancy and load balancing. Also, individual hardware components may be upgraded to faster and more robust equipment as technology advances. In current computing environments, data throughput is limited by the speed of the network infrastructure, and network links can be upgraded as needed.

3. SECURITY ISSUES AND ASSUMPTIONS

The security of this architecture depends on the proper functionality and object reuse characteristics, as well as the security and continued integrity of the individual client workstation's operating system and application software used to initiate data access requests and receive data from different compartments. For example, once data have been retrieved to a shared workstation, the client application and platform must clean up after itself, which includes emptying history and cache files and wiping the application memory and file system, or the next user can obtain access to the data via the temporary files and storage. In a more malicious example, programs running on a client workstation must not capture sensitive data retrieved from a trusted server and forward it to an unauthorized user. Another potential danger of a compromised client is a denial of service attack, where software on the workstation could garble the digital signature on an outgoing request, causing the signature to be unverifiable on a trusted server and legitimate access to be denied. Controls to prevent such accidental or malicious threats to the client workstation exist and are described below.

Therefore, to offer a reasonable level of assurance in this architecture, we must be able to make the following assumptions about client workstations:

- A client is functioning properly, with no accidental or malicious disclosure of sensitive data, data corruption, or denial of service. In addition, a client will not save or transmit protected data without the user's knowledge. Also, an authorized user presented with a sensitivity level will not deliberately capture or forward sensitive data.
- A client runs only as a client, i.e., there are no server services such as an FTP server, file sharing or a telnet server running. This implies that a client is not remotely accessible and users wishing to use a client workstation to access data must have physical access to that machine.

The level of trust required in the client's functioning properly depends on the application and the sensitivity of data being accessed. For many applications, it is sufficient to use a standard commercial operating system because it has been tested extensively by the vendor and through routine use in the field. However, for an application requiring further assurance of the operating system and client workstation functionality, the clients to be used for the application should be certified by a trusted certification agency.

Once we are confident that the client is performing properly, there are several mechanisms available to ensure that the client is not corrupted and that no malicious code can be run, either deliberately or inadvertently. Individual workstations can be booted from a smartcard using the patented Boot Integrity Token System (BITS) (Clark and Hoffman, 1994). The BITS stores a computer's boot sector on a smartcard and requires the smartcard and a password to boot the machine. Once booted from BITS, the computer is assumed to be operating correctly and can run antivirus software and other integrity checks to assure continuing trust in client applications. Strict configuration control can be used to manage legitimate changes to the client.

In addition to reliable software and hardware controls on the client, users must be responsible for protecting data retrieved onto a client workstation. Any data saved to the local client by the user must be removed or protected (e.g., by encryption) before another person, who may not have the same authorizations, uses the workstation. If the client is a mobile workstation such as a laptop computer, the machine's owner must take care to physically protect it.

Enforceable policies must also be in place to prevent users from deliberately or inadvertently circumventing the security inherent in the network configuration. For example, users must not connect workstations directly to the trusted server network domain and bypass the firewall. Some of these threats can be managed by placing tight controls on configuration management or by attaching network sniffers to trusted server domains to monitor for inappropriate network traffic and alert administrators if necessary. Nevertheless, user compliance is vital in this and other compartmented data access models. It can be encouraged through education and administrative or legal remedies and is assumed for the purposes of this discussion.

In these assumptions, the architecture described here is no different from any other implementation of secure compartmented data access. For example, in a paper-based system, high-security controls may be in place to ensure that only authorized users see certain classified documents. However, if an authorized user leaves sensitive material lying around in the open, all security measures may be for naught.

4. COMPARISON WITH RELATED WORK

The work presented in Dalton (1999) uses the secure shell (SSH) remote login protocol to access data on compartmented mode workstation (CMW) hosts from COTS clients over an untrusted network. Although this system requires modifications to SSH, in addition to a specialized operating system on the data "server," it illustrates the power of data accesses from widely available commercial clients such as Windows NT. In this model, servers and clients authenticate to each other before any data or even requests for data are transmitted, whereas in our solution clients generally send digitally signed and encrypted requests through cryptographic gateways before having any assurance of their identity. Receipt notifications and query results returned from the trusted servers are also digitally signed and encrypted. While our architecture can support a mode of operation where all components authenticate themselves to each other before communicating, this requires additional steps, software, and safety measures.

Domain and type enforcement (DTE) (Badger et al., 1995, 1996) also provides user-based access to data. The DTE enforces mandatory access control through operating system modifications to associate types with objects on a system, for example files, and domains with processes on a system. A user chooses a role upon login, which is associated with a domain. A security specification file specifying each domain's access rights to different types is read at system boot into data structures. During system operation, kernel mechanisms check the data structures to verify authorizations for accesses to types from domains. This is analogous to our ACL file described earlier. The DTE protection is extended across networks (Oostendorp et al., 1997) in a manner that protects connection-oriented and transaction-oriented (connectionless) protocols.

Unlike our approach, the DTE research prototype requires significant modifications be made to the operating system and to some standard applications (e.g., login) to provide the mandatory access control mechanisms. Each new operating system release entails porting these changes to the new release and retesting the system. Since our approach uses unmodified COTS products, these updates are unnecessary. The distributed nature of our architecture makes it scalable to greater numbers of users, since redundant components can be added to provide load balancing and fault tolerance. Currently, our approach assumes that objects are usually not created, but instead updates are made to existing databases, etc. Under this assumption, requiring administrative intervention to create and label new objects is acceptable. In cases where new objects are frequently created, our approach could benefit from the scalability provided by DTE's implicit attributes, in which type labels are implicitly inherited from parent objects until over-ridden by an explicit label specified in the security specification.

Also, DTE performs user authentication only once rather than for each transaction. At login, DTE can achieve any desired level of assurance in user authentication, from simply requiring a password to using token-based or biometric authentication. However, once the user has logged in, he or she is authenticated and trusted by the operating system for all further actions performed during a session. Our approach authenticates and protects each data request and submission.

5. FUTURE DIRECTIONS

Our secure compartmented data access model works well for submissions to and retrievals from files and databases that are already in place. However, creating or modifying data structures such as database tables or subdirectories requires administrator intervention and is not very scalable. Planned enhancements include revisions to the way access to resources is controlled through ACLs. We anticipate that making access control inherited from parent to child data structures will make data resource creation less cumbersome while retaining a high level of security. For example, a newly created subdirectory would inherit its authorized users from its parent directory. In conjunction with this modification, we will maintain the capability to over-ride inheritance, similar to the DTE model discussed earlier.

Besides making the system more powerful, inheritance of access rights increases its usability for system administrators. Other improvements in usability that are being investigated for future work include tools for secure remote user registration and high entropy key generation with a secure web interface and remotely existing issuer credentials.

6. CONCLUSION

Other systems exist that provide controlled access to sensitive, compartmented data. However, the architecture described here enables electronic business on a broad scale by using scalable and maintainable COTS-based server solutions with low or no-cost clients. Our approach implements a distributed TCB using COTS hardware and software components. Compartmentalized sensitive data are remotely accessible over an untrusted wide area network but cannot be directly accessed by users, thereby providing high-assurance protection of the data. We provide a trusted path for the data from its protected storage to authorized users. By allowing cryptographic enhancements to be transparently applied to individual transactions, usability of the system is increased in comparison to other secure systems and even relatively insecure systems—for example, those that merely require and rely on the use of user passwords. Due to its transaction-based nature, the presented solution also supports a secure audit capability and resource-based authorization model. Finally, this architecture has been deployed and tested in multiple production environments with different assurance levels, where it has successfully met the challenges of providing secure, controlled access to sensitive compartmented data in a low-cost and usable manner.

Secure compartmentalized access in a networked environment is not only possible but it can be accomplished using standard, relatively inexpensive equipment and software. The use of COTS products and standard algorithms make for a flexible and scalable solution. The described architecture provides embedded strong user-based authentication and encryption for all accesses to trusted server resources over the wide area network. Although the architecture is subject to the same physical and administrative requirements as non-networked systems and special purpose secure data access systems, it provides efficient technological mechanisms to help monitor and enforce diverse security policies.

REFERENCES

Badger, L., Sterne, D. F., Sherman, D. L., Walker, K. M., Haghighat, S. A. (1996). A domain and type enforcement Unix prototype. *USENIX Computing Systems*. Vol. 9. Cambridge, MA.

Badger, L., Sterne, D. F., Sherman, D. L., Walker, K. M., Haghighat, S. A. (1995). Practical domain and type enforcement for Unix. Proceedings of the 1995 IEEE Symposium on Security and Privacy. Oakland, CA, May.

Choo, T.-H. (1999). *Vaulted VPN: Compartmented Virtual Private Networks on Trusted Operating Systems*. HPL-1999-44, Extended Enterprise Laboratory, HP Laboratories Bristol, Mar.

Clark, P. C., Hoffman, L. J. (1994). BITS: a smartcard protected operating system. *Communications of the ACM* 37(11).

Dalton, C. I. (1999). *Strongly Authenticated and Encrtyped Multi-level Access to CMW Systems over Insecure Networks using the SSH Protocol*. HPL-99-98 (R.1), Extended Enterprise Laboratory, HP Laboratories Bristol, Feb.

National Computer Security Center (1985). Department of Defense Trusted Computer System Evaluation Criteria. DoD 5200. 28-STD, Dec.

National Computer Security Center (1993). Trusted Network Interpretation. NCSC-tg-005 (Rainbow Series), Nov.

Oostendorp, K., Badger, L., Vance, C., Morrison, W., Sherman, D., Sterne, D. (1997). Domain and type enforcement firewalls. Proceedings of the 13th Annual Computer Security Applications Conference. San Diego, CA, Dec.

6

Internet Security

MATT BISHOP, STEVEN CHEUNG,
JEREMY FRANK, JAMES HOAGLAND,
STEVEN SAMORODIN and
CHRIS WEE

Department of Computer Science, University of
California at Davis, Davis, CA, USA

1. INTRODUCTION

The previous article has explored issues of who commits crimes using the Internet, and why. One obvious question is whether the Internet provides security mechanisms and protocols that can prevent breaches of security, such as those leading to crime. Could not an infrastructure be designed to support detection and analysis of breaches of security, and to trace them to their source?

Manuscript submitted for publication 2003.

The Internet is a collection of networks, owned and operated by many organizations all over the world. The common element across all the networks is the use of Internet protocols, specifically IP, the Internet Protocol (which underlies the other protocols) and its cousins (e.g., TCP, UDP, and DNS). Since it is a global phenomenon, there is no uniform cultural, legal, or legislative basis for addressing misconduct on the Internet. Several issues fundamental to the origins and structure of the Internet and its protocols and to computer security in general, limit our abilities to prevent intrusions, to trace intruders, or even detect intrusions. Understanding these issues in the context of the Internet will answer our questions.

The primary issue is trust. When we glean information from the Internet protocols to determine whether to allow access, or to identify the origin of a connection, we trust the network services to provide correct information. If they do not, attackers can exploit this failure to gain access to a protected system. So the trustworthiness of network support services directly affects the security of the Internet. A related issue is the strength of the implementations of the protocols. Failures in software engineering are common, and implementations rarely perform exactly according to specification—more precisely, while they may implement the specifications correctly, they usually do not handle errors correctly, or the interaction of the program with other programs or users introduces security vulnerabilities. An intruder can often exploit these flaws to access data or systems illicitly; these flaws also hamper detection and recovery. These vulnerabilities in implementation, as well as design, also directly affect the security of the Internet. This paper explores these issues.

1.1. Policy

Information security may be defined as the goals and practices that regulate data sharing and data exchange. Just as a legal system regulates the activities and interactions of citizens in a society, information security regulates and facilitates the shared use of a system, its resources and its data. Since the Internet is a shared global resource, users (“netizens”) or their aggregate institutions must be held accountable for their activities on the net. Typical policy objectives include protecting confidentiality, preventing unauthorized modification (integrity), and preserving availability of resources according to the expectations and needs of the users. The data include not only data that is stored on hosts connected to the network but also communications that traverse the network such as e-mail or World Wide Web traffic. The Internet is too large and too diverse to have a single policy, and has no central authority to regulate behavior. Most organizations connected to the Internet have their own security policies, which vary widely in statement and objectives.

Security policies, like laws, are difficult to express precisely or state formally. A formal security policy may consist of a model of the system as a collection of states and operations and a set of constraints upon the states and its operation. In current practice, a system security policy is stated in ordinary language; the imprecision of such a statement makes translation to another form mandatory if the policy is to be enforced automatically. The imprecision, however, is not the only problem; typically, configuring security mechanisms to enforce the policy is error prone and incorrect configurations allow violations of security policies.

Violations of security policies are called attacks or intrusions, even if committed accidentally. Intrusions may be classified as masquerading (impersonating a user or resource), unauthorized use of resources, unauthorized disclosure of information (violating confidentiality), unauthorized alteration of information (violating integrity), repudiation of actions, and denial of service. The security policy distinguishes authorized behavior from unauthorized behavior. For example, the sharing of data (disclosure) between users is only considered a violation if the sharing is not authorized. Thus, the security policy is the standard against which user activities are judged benign or malicious.

The role of policy in determining what constitutes an intrusion, and how serious that intrusion is, guides the development of detection, assessment, and prevention mechanisms. These mechanisms rely upon the logging mechanisms embedded in the systems on the Internet, and the Internet infrastructure. The logs provide invaluable information for intrusion detection and analysis; indeed, they form the basis for all post-mortem analysis. Indirectly, the policy determines what to log, how the desired level of logging impacts system performance, and how to analyze the resulting logs. Their critical role in Internet security makes them an important topic for our third section. The fourth section discusses how the policy, and these associated mechanisms, guide detection, assessment and recovery from attacks, as well as prevent attacks.

Policies fall into two classes: military or governmental, and commercial. A military policy emphasizes confidentiality; a commercial policy, integrity. For example, the Navy needs to keep confidential the date on which Sam's troop ship is sailing; if that is changed on a computer, sufficient redundancy exists that the correct date can be determined. But a bank needs to prevent unauthorized changes to Robin's accounts. Disclosing Robin's balance may be embarrassing and actionable, but the financial stability of the bank will not be threatened, and the bank's very existence depends upon depositors trusting the bank to prevent those balances from being illicitly changed.

A very common conflict of policies arises when one company acquires another. If the first company maintains a closed security policy (which denies access by outsiders) and the acquired company has an open security policy (encouraging access by outsiders), what will the policy of the conglomerate be?

2. PROTECTION

Policy enforcement, like law enforcement, forces users' activities to conform to the policy by preventing unauthorized behavior and deterring malicious activities. Policy enforcement mechanisms are broadly classified as protection or audit. Protection seeks to prevent malicious behavior while audit aims to detect and deter malicious activity. Protection mechanisms include access controls, cryptography, authentication, and redundancy. Audit mechanisms are investigative tools that examine system activity, or records of system behavior (called "audit logs") to detect malicious behavior. Audit is also used to review the effectiveness of access controls. The combination of protection and audit mechanisms enforces the security policy.

However, policies cannot be enforced exactly due to limits of the technology used for the system's protection mechanisms. The limits affect the granularity, administration, configuration, and operation of the mechanisms. For example, the file protection mechanism of the UNIX operating system is suitable to limit access to files, but not to prevent users from copying files they can read. In addition to the technological gap, there is a sociotechnical gap between social policies and information security policies that arises when social norms are not consistent with information practices. For example, social policies may distinguish between unintentional mistakes and malicious actions, and even tolerate unintentional blunders. But information security policies cannot distinguish between them because computers cannot recognize a user's intent. There is even a social gap between social policies and actual user behavior. Authorized users can easily deviate from social norms and commit abuses.

Protection is the primary defense against malicious behavior, restricting user activities to those allowed by the policy. For instance, a system may be designed to accommodate only one user at a time to satisfy a policy requirement for isolation and privacy. When a user wishes to access a protected object, the system's access controls determine whether the security policy authorizes that user to access that object. If the access is authorized, the user gains access; if the access is unauthorized, the user is denied access and an error code may be returned to the user. The access decision is usually made at run-time, at the beginning of each access. However, on some systems, only the first access is checked; on these systems, the level of granularity inhibits the enforcement of policies that require checking every access to an object.

The information security life cycle is an iterative process of identifying users' security expectations (policy), enforcing the policy, then re-assessing the system in light of policy violations or intrusions, and repeating the cycle by modifying the original policy specification. The information security life cycle is similar to the software engineering spiral cycles. The activities (in italics) that are performed at each stage (in the circle) of the information security life cycle are illustrated in Figure 1 .

Like laws in a legal system, security policies must continually evolve to address the users' and organizations' expectations more accurately and precisely, and to accommodate changes in the users' and organizations' security requirements. Also, like crime prevention, protection and investigations must keep pace with new policy violations, and with the users' new work patterns and their new applications. During the recovery step of each iteration in the life cycle, both policy and protection mechanisms are refined to address new attacks, to close vulnerabilities, and to update the policy to accommodate new user and organization requirements. This is analogous to a legal system that evolves by court decisions that interpret and clarify laws, by upholding or prohibiting enforcement procedures, by amending existing laws, and by interpreting new laws. Flexibility to change as the policy evolves is one of the features of audit-based intrusion detection systems.

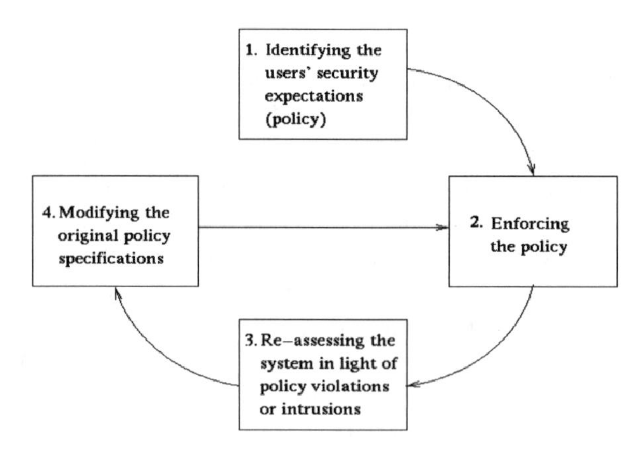

Figure 1 Protection and audit in the information security life cycle.

Robin and Sam are running a company to design Web pages. They started off using a single computer to design and display their products. The security policy required them to allow anyone read access to their displayed Web pages, but not access to the pages under construction. So their mechanisms must restrict access to the Web pages on display. Later, their company expanded to allow them to buy another computer. They moved their designing to the new system, and simply disallowed all Web access to that system. Their original security policy remained the same, but the mechanisms to enforce it were used very differently.

3. TRUST AND VULNERABILITIES

Assume an attacker, called the Adversary, wishes to violate a target's security policy and do harm or mischief to it. The Adversary tries to attain this goal by exploiting vulnerabilities, which are avenues to violating the security policy. The Adversary identifies one or more vulnerabilities and develops a sequence of actions, called an *attack*, which exploits the vulnerability, resulting either in damage to the target or rendering the target subject to further action which results in damage. Without any vulnerabilities, the attacker would fail to achieve the goal.

Vulnerabilities in systems arise because the designers and implementers make assumptions about the environment in which the system will be used, the quality of data on which it will work, and the use to which it will be put. They make tradeoffs in the name of development cost and time, eventual price, ease of use, efficiency and numerous other qualities. For example, components of systems may perform detailed checks on the information they receive from other components, but this trades quality of service for increased computation time.

The assumptions come from experience, beliefs about the environment, in which the system works, and laws and cultural customs. For example, when designing the cryptographic protocols used by Netscape, the designers assumed that there were no constraints on key length. The implementers realized that United States law restricted the maximum key length, and used a (legal) length of 40 bits. Trading exportability for security enabled a group of graduate students from the University of California at Berkeley to break a message sent using the security-oriented communications protocol in a matter of hours by testing all of the possible keys. Here, the tradeoff resulted in a level of security inadequate for most uses.

Vulnerabilities also arise when systems are not used according to the assumptions under which they were built. As an example, firewalls are systems that do not forward some network communications as dictated by a site security policy. If a site wishes to prevent access to a World Wide Web server from the Internet, the firewall would be configured to block communications on port 80 (which the World Wide Web protocol, HTTP, uses). However, if someone on the other side of the firewall ran a World Wide Web server that accepted connections on port 23, the firewall would allow communications to that server through, violating the site security policy (and the assumption that all HTTP communications will go through the firewall at port 80).

Another example of incorrect assumptions occurs when a program fails to verify that the input data are valid. One mail-receiving program used the *ident* protocol to attempt to verify the identity of the sender of the letter. The *ident* protocol transmits a short message containing the name of the user of a network connection. The mail-receiving program assumed this message would be fewer than 512 characters long. An attacker could break into the computer on which the mail-receiving program ran by sending a letter, and responding to the *ident* request with a carefully crafted message that was over 512 characters long. The mail-receiving program trusted the input from the *ident* server—and that input caused a violation of the security policy by providing unrestricted access to the target system.

Suppose Sam sends an ident request to Robin's system. This raises several questions of trust. How does Sam know that Robin's system will return the correct information? Will

Sam interpret the returned user identifier in light of Robin's people rather than his (as the user "bishop" may refer to Leonard Bishop on Sam's system and Matt Bishop on Robin's)?

Vulnerabilities may also arise as a result of flaws in the implementation of the system. Even if the program design conforms to the security policy, implementation errors can cause violations, and hence create vulnerabilities in the system. Writing bug-free programs requires detailed analysis, careful coding, and thorough testing, but programmers often work under constraints (such as lack of time, money, or other resources) inhibiting these activities. Vulnerabilities can be very subtle and can exist in systems for years before being noticed; further, the conditions under which they can be exploited may be very brief and transient. The JAVA system, used to provide downloadable executable programs called *applets*, is a good example of a system with an implementation flaw. Although the designers attempted to limit the dangers to the downloading system by restricting the allowed actions to a very limited domain, a number of implementation flaws allowed the applets to take actions inconsistent with those restrictions

Initial implementations of Java suffered from many of the errors described above. Some of the printing functions did not check bounds, so attackers could send long messages to Robin and Sam's system, and cause the Web server to execute any instructions in those messages. (This was why Sam and Robin moved their Web designing to a separate system.) Java constrains the applets to connect back to the system from which the applet was downloaded, and disallows connection to any other system. The systems are named in (by) the applet, but are compared using network addresses. The problem is that the applet can ask to connect to *any* system on the Internet. A request to obtain the network address from the name is sent out over the Internet, and another system (the domain name server) returns the network address. If that domain name server is corrupt, it can lie and say the host to be connected to has the same address as the host from which the applet came. The implementers trusted that this look-up would be correct and reliable—and in the face of an attacker, it need not be. This has also been fixed by referring to all systems by network address and not name.

Configuration errors also cause vulnerabilities. For example, most World Wide Web servers allow system administrators to restrict access to Web pages based upon the address of the client; should the system administrator mistype an address, or fail to restrict sensitive pages, the company security policy can be violated. Any time a system administrator or user must configure a security-related program, a potential vulnerability exists.

Hardware vulnerabilities are usually more subtle, but can be exploited. For example, researchers have artificially injected faults into smart cards to determine the cryptographic keys used in the encryption codes. The "burning" of keys into hardware was supposed to protect them; but it did not protect them well enough.

Vulnerabilities are not confined to end systems; the computers, protocols, software, and hardware making up the Internet have flaws too. Consider a router, which is a computer designed to forward packets to other routers as those packets go through the Internet to their destination computers. A router uses a routing table to determine the path along which the packet will be forwarded toward its destination. Periodically routers update each other's tables, thereby allowing a dynamically reconfigurable network. If through design or error a router were to announce that it was a distance of zero from all other routers. Then all other routers would send the misconfigured router all of their packets. It would try to reroute them, but all routes would lead back to it, so the packets would never reach their destination. This causes a denial of service attack.

The central theme of vulnerabilities is one of *trust*. The dictionary defines "trust" as "confidence in some quality, feature or attribute of a thing". The designers and engineers trust a system will be used in a particular way, under particular conditions. When systems are designed in pieces, the design teams trust that the other teams did their jobs correctly so that pieces fit together with a minimum of difficulty. When systems are composed of many subsystems, each subsystem is trusted to do its job correctly. The designers of a program trust that the programmers do not introduce errors into programs. When consumers purchase a system to use, they trust it will perform as specified. When this trust is misplaced, vulnerabilities result.

Denning lists several ways in which trust can be misplaced. The following list applies some of those ways to a broader context.

- A system might not be operated in the configurations in which it was evaluated.
- A system might be evaluated using a model that fails to adequately capture the true complexity of the system; as a result, important system behavior is overlooked.
- Objective models might not map directly to systems, so objective criteria of trust might not apply to real systems.
- The systems might have too many controls, too much documentation, and confusing user interfaces.
- Developers and implementers might add errors or bugs to the system.
- The "composition" of secure systems is not necessarily secure; the nature of the composition and the requirements of the composition must be clearly stated, and all too often this issue is ignored.

Some of these ideas overlap. The difference between a "bug" and a "configuration error" is a matter of context; a user can reconfigure a system to change its behavior, while a programmer can rewrite a piece of software to change its behavior. Systems are frequently analyzed using models; sometimes the models are created from the system, while other times models are imposed on systems. With respect to the Internet, which is composed of many smaller systems and protocols and is itself a large system, it is impossible to analyze fully all aspects of a system that complex, and so users implicitly trust that the Internet will function correctly when used.

The vast scope of the Internet as an enterprise means that trust is essential to produce working Internet services. Consider electronic mail as an example. Suppose Robin in Seattle wants to send a love letter via electronic mail to Sam in Tampa (see Figure 2) Robin types the letter using a mail program on a computer, and sends it. Robin implicitly (or explicitly) trusts that the following chain of events occurs:

- The mail message contains the letter Robin typed and not some other letter entirely.
- The mail program correctly sends the message via the local network that Robin uses to the next network.
- The message is sent through the interconnected networks making up the Internet, to Sam's computer. This involves routers, which select paths through the Internet for messages.
- The destination computer's mail handling program must successfully receive and store the message, and notify Sam that a message has arrived.
- Finally, Sam must read the message successfully using a mail reading program.

This requires the successful operation of multiple pieces of hardware, including computers and dedicated routers, and

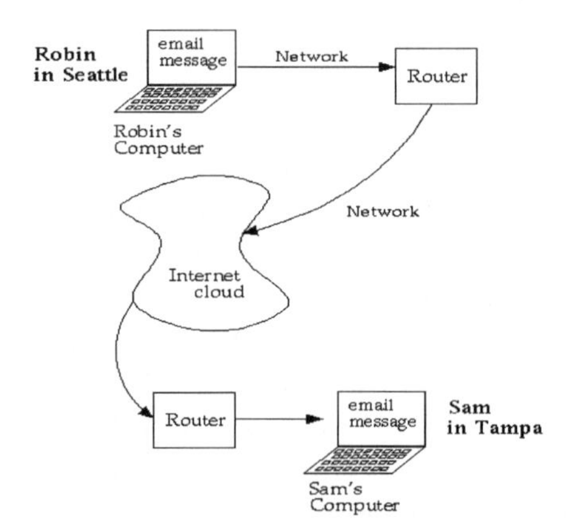

Figure 2 An example connection between Robin in Seattle and Sam in Tampa.

the transport medium, be it twisted pair, fiber optic cable, or satellite link. In addition, this process requires the successful operation of numerous pieces of software on these hardware platforms, including the mail programs, the operating systems, and the software that implements the transportation of messages. The number of distinct components can become quite large, perhaps over one hundred for a message traveling across the country. They must all interact correctly to guarantee that electronic mail is delivered correctly.

As an example in which misplaced trust leads to vulnerabilities, suppose that the Adversary is competing with Robin for Sam's affections, and wants to alter Robin's mail. As mentioned above, Robin trusts that the mail system delivers the message to Sam. But messages traveling over the Internet can be modified en route, so the message received is not the message sent. To do this, the Adversary instructs the routers or mail program to forward all messages to some intermediate system. The Adversary can then read messages on this intermediate site, change their contents, and forward them to the original destination. This is known as a "man in the middle" attack because the Adversary gets "in the middle" of the communication and changes it.

As the Adversary can exploit the vulnerabilities in the Internet systems and change the message to read something else entirely, Robin's trust that the Internet will deliver the message to Sam is misplaced. Hence, Robin is vulnerable to an attack resulting in a different message being delivered. To move the premise into the business world, if Robin is an executive sending mail on corporate strategy to Sam, then the impact of this vulnerability can be disastrous.

Cryptography, often seen as the answer to this problem, is merely a part of the solution because of a simple yet fundamental trust problem: how do you distribute cryptographic keys? Public key cryptosystems provide each user a public key, which is to be widely distributed, and a private key known only to that user. If Robin wants to send Sam mail, Robin enciphers a message using Sam's public key and sends the enciphered message to Sam (see Figure 3). Sam can decipher the message using Sam's private key, but without Sam's

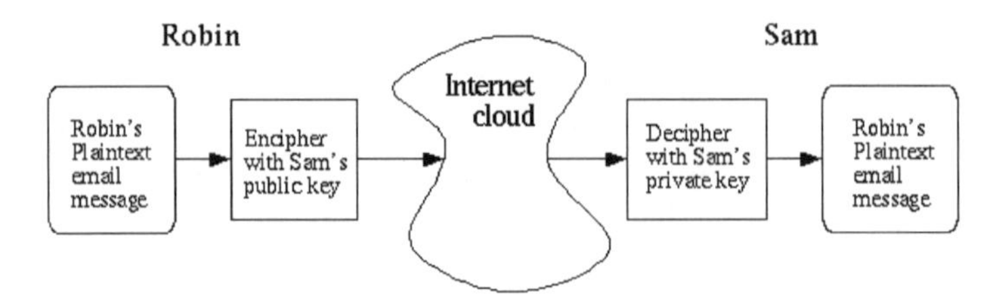

Figure 3 Confidentiality using public-key cryptography.

private key the Adversary cannot decipher the message. Unfortunately, the Adversary may be able to convince Robin that the Adversary's public key is in fact Sam's public key. As a result, when Robin enciphers the message using the wrong key and sends that message to Sam, the Adversary can intercept the message and decipher it. This example shows the pervasiveness of the trust issue; Robin must trust that Sam's public key really belongs to Sam, not someone else. Other nuances of this problem also affect trust.

The situation becomes even more complicated with the World Wide Web. Suppose Robin uses a Web browser to view a Web site in Germany. The Web page has a link on it that says: "Click here to view a cool graphic image." Robin clicks on the link, but in addition to showing the graphic, the link causes Robin's browser to download a program that scans Robin's system for vulnerabilities and mails a report to the Adversary.

Robin trusted the information on the Web page in two ways; first, that an image would be loaded, and second, that *nothing else would be loaded*. This second element is violated by computer programs containing Trojan horses and viruses as well; users of PCs spread viruses by trusting that their programs do only what they are documented to do, and therefore they fail to take the necessary precautions.

Users of the Internet must be aware of the components that they trust to work properly. These include the (hardware and software) infrastructure components of the Internet as well as the computer and network hardware and all the software on the computers involved in the communication. They also trust the design of the protocols that the hardware and software support. This caveat also goes for designers of Internet-based services.

4. AUDIT

Because a policy may be specified in English, and hence not provide clear and unambiguous rules of what is and is not allowed, the violations of policy might not be automatically detected. The same social, sociotechnical, and technological gaps that make policy specification imprecise preclude the unambiguous translation of the definition of intrusions into effective and efficient protection and audit mechanisms.

The ambiguity of the intrusions notwithstanding, detecting attacks affects the user's beliefs about the security of the system. Auditing is the practice of enforcing five control objectives to detect and deter attacks:

- Trace accountability for accesses of objects to individuals who may then be held responsible for their actions.
- Verify the effectiveness of system protection mechanisms and access controls.
- Record attempts to bypass protection mechanisms.
- Detect uses of privilege greater than or inappropriate for the role of the user.
- Deter perpetrators and assure the user that penetrations are recorded and discovered.

These goals do not dictate a particular model of audit, nor do they indicate how to perform the auditing process to meet these objectives. Currently, auditing consists of various ad hoc practices that accomplish many of these control objectives.

Auditing a system requires that user activities and protection-related information (called *audit events*) be recorded. A *log* is a collection of audit events, typically arranged in chronological order that represents the history of the system; each audit event in the log represents a change in the security-relevant state of the system. These logs can be voluminous because of the complexity of modern systems and the inability to target specific actions; the logs are often so large, in fact, that human analysis is quite time-consuming. The logs should be analyzed using tools to cull the entries of interest. The development of these automated audit tools is hampered by the lack of standard audit formats with standard audit log semantics, and the lack of mechanized representations of security policies. Some tools exist to aid analysis, but they are difficult to use, so logs are either manually inspected (often in a cursory manner), possibly using some audit browsing tools that employ clustering algorithms, or not reviewed at all.

When the log is reviewed, the users' activities are compared to the policy, and the auditor reports any policy violations. An auditor can also use the log to examine the effectiveness of existing protection mechanisms and to detect attempts to bypass the protection or attack the system. The identities of those responsible for attempts to violate the policy sometimes can be traced in the history of events. On networked computers, logs from several hosts may be required to trace the user, who may be located on a host remote to the host where the violation was attempted. Logs may be used as evidence when prosecution of the perpetrators is warranted.

The basic design philosophy of the Internet places the resources and capabilities at the host end points to keep the infrastructure simple, flexible, and robust. The disadvantage of this design philosophy is that the Internet Protocol offers only to make the best effort to deliver messages, rather than guaranteeing their delivery. As a result, logging on the Internet is usually a function of implementation and not a requirement of the protocols. Packet monitoring, also known as *sniffing*, logs packets on a broadcast-type local area network (LAN). Sniffing also records packets sent over a point-to-point network link. Depending on the amount of traffic on the LAN or link, sniffing can be both CPU and storage intensive. Sometimes, only the header portion of packets is logged; the data in the packet are ignored. Deducing user behavior and applications' higher-level actions from information obtained by sniffing requires many translations between the network data and the application actions, and involve many assumptions.

Most World Wide Web servers use a standard audit log format for web servers, allowing audit tools to be developed for a wide range of servers. Also, electronic mail often has the name of each host, and some additional information, placed in the headers of the message as the mail moves from host to host. These headers constitute a mini-log of locations and actions that can be analyzed to diagnose problems or to trace the route of the message.

Although protection mechanisms are designed to prevent violations of the security policy, the most successful deterrent is often the specter of accountability and the attacker's fear of being discovered. Audit mechanisms, then, may be thought of as a secondary defense against attacks. This use of audit is primarily referred to as intrusion detection.

Auditing provides different levels of information about a system. Three very popular (and free) tools for auditing UNIX systems are SATAN, COPS, and tripwire.

SATAN is a World Wide Web-based program that analyzes a system (or systems) for several well-known vulnerabilities exploitable only through network attack. It provides a Web browser interface, and allows one to scan multiple systems simply by clicking on one button. The browser presents a report outlining the vulnerabilities, and provides tutorials on the causes of each, how to close (or ameliorate) the security flaw, and where to get more information (such as any related CERT advisories).

System integrity relates to Internet security through the software implementing network servers. Tripwire is a file system integrity-checking program that computes a mathematical function ("hash") of the contents of the file. These hashes and the names of the corresponding files are stored in a file. Periodically, a system administrator reruns tripwire, and compares the results with the results of the original run. If any of the hashes differ, the corresponding file has been altered.

COPS examines the contents of files and directories and determines if either their contents or settings threaten system security. For example, on Sam's UNIX system, the contents of a file will determine if Robin needs to supply a password to use Sam's system. This poses a security problem at many sites: anyone who can obtain Robin's login name also obtains access to Sam's system. Tripwire will not detect this problem, as tripwire simply looks for files that changed—and the access control file does not change. COPS will scan the file, and report that Robin does not need a password to log in, as it analyzes the contents of the file.

5. INTRUSION DETECTION

The IDS obtains information from the system and logs, processes it, and reports problems to the auditor. It may also use that information (feedback) to control the generation of audit data and future processing of logs.

Intrusions can be detected either by humans analyzing logs or by automated tools that detect certain specific actions or system characteristics. The automated tools are known as *intrusion detection systems* (IDSes). Automated methods offer the advantages of processing much data quickly, and more efficiently, than humans could. The data come from either logs or from the current state of the system (see Figure 4).

Human analysis of logs requires looking at all or parts of the logs for a system and attempting to uncover suspicious

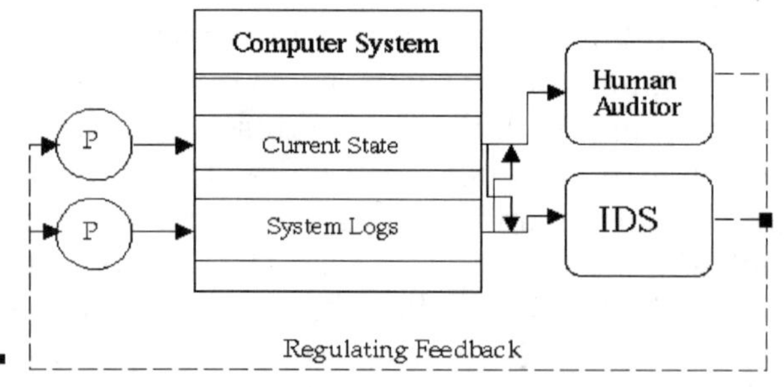

Figure 4 Intrusion detection.

behavior. However the audit data may be at such a low level that events indicating an intrusion or attack may not be detectable as such, so detecting attacks may require correlating different sets of audit data, possibly gleaned from multiple logs. These data may span days or weeks and is often voluminous, raising the issues discussed earlier. Also, if the amount of activity on a system continues to increase, the amount of audit data also increases, and at some point exceeds any human's abilities to analyze effectively. Another hindrance is that the person conducting the analysis must have expertise on the type of system, and on the particular host, being audited because of the need to understand the relevant parts of the system design and configuration. To detect security violations reliably, the human analyst must remain alert, and not be distracted by false alarms or other disturbances.

J.P. Anderson made the first serious investigation of employing computers to detect security violations of computer systems. Modern computers have a capacity to analyze large amounts of data accurately, providing they are programmed to do the correct analysis. However, they must be told what to look for. Three general approaches are used to detect an intrusion: anomaly detection, misuse detection, and specification-based detection.

5.1. Anomaly Detection

The anomaly detection approach compares current behavior to historical behavior and presumes that deviations are symptoms of intrusions. Unauthorized people using a valid account can be detected by observing user actions, the assumption being the intruder will execute different commands than the authorized user, thereby creating a statistical anomaly. Similar reasoning suggests that a program altered to violate the security policy will behave differently than the unaltered version of the program.

An IDS using anomaly detection must be initialized or *trained* to know the expected behavior of the users. This profile is called a *normalcy profile*. This profile is built using statistical analysis and logical rules. After a normalcy profile is established, the system is monitored and the IDS compares user activity to the user's normalcy profile. If some activity deviates significantly from the profile, then the IDS flags it as anomalous and as a possible intrusion. However, false positives arise because abnormal behavior is not necessarily an attack. False negatives also arise when intrusions closely resemble normal behavior. Finally, retraining the detection system and establishing the right time period over which to analyze the behavior affects the results of the IDS.

One anomaly detection system observes the interaction between a program and the operating system, and builds normalcy profiles of the short sequences of system calls normally made. Activity outside this is presumed to be part of an intrusion. For example, if the Adversary tried to exploit a vulnerability in which unusual input caused the mail receiving program to execute unexpected commands, these would be detected as anomalous and a warning given.

5.2. Misuse Detection

Unlike anomaly detection, which learns what characterizes an attack by observing the behavior of the system, the misuse detection method requires a priori specifications of attacks. No learning takes place. The system is monitored and if any behavior is observed that matches any of the provided attack patterns, then the IDS warns of an attack. The techniques used to describe the attacks vary. One technique is to list events expected to be logged during the attack. The Graph-based Intrusion Detection System (GrIDS) uses a set of rules that describes how to construct (mathematical) graphs based on network and host activity. These rules also describe at what point the graph is considered to represent an attack.

To see how misuse detection works, return to the example of an attacker giving unusual input to the mail-receiving program. An expert could write a rule set describing the behavior that would indicate a violation of policy. For example, if the expected set of system calls were "read-input; write-file" but unusual input caused the set of system calls to be "read-input; spawn-subprocess; overlay-program", the last sequence would be placed in the ruleset as indicating an attack. Then when the Adversary tried to intrude using that technique, the misuse detection program would detect the attempted violation of security.

The misuse detection method can be highly accurate, but, unlike anomaly detection, cannot detect attacks that fall outside of its provided list. In addition, it depends upon an expert being able to specify rules describing violations of security.

5.3. Specification-Based Detection

Misuse detection catches security breaches by describing the behavior of attacks. Specification-based detection catches breaches by describing what behavior is expected from the system itself. Further, with this technique, there are no false alarms, provided the system has been specified accurately. This method detects attacks based on an abstract specification of what the correct behavior of the system is.

The first step is to specify formally how the system should behave in all cases. The system is monitored and all actions compared against the specification. Behavior is flagged as a security violation if it is observed to fall outside what is specified as correct for the system. For example, if there is a policy that all hosts on a network must use a particular proxy host for Web traffic, then the specification might be that the proxy host can connect through the Web to anywhere, but all other hosts' Web connections must be to the proxy. Then if Web traffic from a host other than the proxy to a host other that the proxy is observed, the security policy is violated.

One approach specifies privileged programs in terms of a program policy specification language. This language indicates under what conditions particular system calls may be made and requires knowledge about privileged programs, what system calls they use, and what directories they access. Depending on the system being specified and the specification language used, creating these specifications may require expertise, skill, and some time. However, some of this might be automated through program analysis. Further, this approach does not eliminate false negatives, as the specifications may not be complete.

There is no one best approach to detecting intrusions. In practice, the particular combination of approaches used is tailored to the specific needs of the system and organization.

Several companies and research groups have developed intrusion detection systems. The UC Davis Computer Security Laboratory is designing and developing one such tool, called GrIDS, which will monitor both systems and network traffic, looking for actions indicating misuse. It will also coordinate analyses with other GrIDS systems, and supports analysis of distributed attacks, even when the attack is spread over a large number of systems. Nonresearch systems, such as NetStalker, are less ambitious but are currently deployed, and can detect attacks against systems by comparing system actions to known exploitations of vulnerabilities. NIDES uses the anomaly approach by building a database of statistical user profiles, and looking for deviations from that profile.

6. RESPONSE

Once a violation of the security policy has been detected, several responses are possible, particularly if the attack has been detected in its early stages. The simplest response it to alert humans to the violation using electronic mail, beepers, alert windows, or other means. An automated detection system might be configured to respond autonomously to any violations of policy. Regardless of how the system is activated, different response options are available. The response selected depends on the degree of confidence that an attack is actually under way, and upon the nature and severity of the attack.

The first response to a reported attack is to gather the information needed to analyze the violation of policy and determine how to respond further. For example, additional auditing may be enabled, possibly only for those users involved in the violation, or possibly for the entire system.

The latter is useful if the extent or nature of the violation of policy is not fully understood. In addition, the system can fool the attacker by returning misleading or incorrect information to the attacker; the attacker can even be lured to a system specially designed to monitor intruders.

Another common response to a policy violation is to determine the accountable parties. After that, legal action might be taken, or more direct responses (such as blocking further connections from the attacker's site or logging the attacker off) may be appropriate. However, determining whom to hold accountable can be very difficult as the Internet protocols do not associate users with connections, the attack might be laundered through multiple stolen accounts, and might cross-multiple administrative domains. No infrastructure support exists to trace attacks laundered in this way. However, other means can be used to track intruders.

7. ASSESSMENT

When a policy violation is detected, one response is to determine what damage was done; the system attacked needs to be analyzed. This analysis is the assessment of the system.

Assessment has two goals:

- identify the vulnerability or vulnerabilities exploited in the attack and the source of the attack, if possible;
- identify the damage to the affected systems.

Knowing the vulnerabilities exploited can stop on-going attacks, take steps to close the security vulnerabilities to prevent future attacks, or determine what to monitor if the vulnerability cannot be fixed. Identifying the damage indicates what must be changed to bring the affected systems back to a previous uncompromised state, and to prevent future attacks.

The first goal is closely related to detection. As discussed above, some anomaly detection systems detect for deviation of users' behaviors from an established profile. However, those detection systems can only indicate that a user may be an attacker, but not what vulnerabilities were exploited to violate the security policy. Misuse detection systems detect exploitation of known vulnerabilities, but may give us only a partial set of the vulnerabilities exploited, because the activities that trigger the IDS may not be the root cause of an attack. An attacker may use other means to violate the policy initially, but that first violation may not be detected. The attacker may then commit additional violations, based in part on the initial one, and those may be the violations reported. Thus, additional work may be needed to uncover the vulnerabilities exploited.

Locating the source of the attack means identifying the machine from which the attack is launched and the person who launches the attack. Tracing an attack that involves multiple intermediate machines is a hard problem, as discussed above. Identifying the source may suggest countermeasures to stop on-going attacks as well as prevent future attacks (such as refusing connections from the attacker's system).

The second goal of assessment is to determine the damages to the affected systems. The damage involved arises from the nature of the policy violation; examples of damage are unauthorized disclosure of confidential information, causing the target systems to be unavailable, and unauthorized changes to data or programs.

Successful assessment depends upon the integrity of the audit data and the analysis programs used for the assessment. This is a trust issue. A sophisticated attacker may disable or modify the analysis programs or tamper with the audit data to hide the attack. Thus extra resources are needed to secure those data and programs. For example, audit data may be written to write-only or append-only devices, and analysis programs may be put on a dedicated machine that does not have ordinary user accounts, does not have any network connections, and uses the vendor's distribution of the operating system.

Recall that Robin wanted to send an electronic message to Sam. Suppose the Adversary successfully compromises Robin's machine. To eavesdrop on the communication between Robin and Sam, the Adversary replaces Robin's mailer with a mischievous mailer that copies the messages sent between Robin and Sam to Adversary, in addition to actually sending the messages. As a more complex example, suppose the Adversary exploits the trust between Robin's machine and a web server machine that allows one to logon from one machine to another without supplying a password. The Adversary modifies Robin's personal web page so it now contains malicious Java code that directs any browser reading that web page to attack other hosts. Making the web page an attack launch pad may potentially introduce liabilities to Robin, without Robin's knowledge of what happened.

Assessment can be approached using event-based or state-based analysis. Event-based analysis requires tracking down the causal relationships among events. This approach depends on logging system events as discussed above. For example, the UNIX operating system associates with each process a process id, the parent process id, and the user that starts the process. Moreover, the process id's are recorded with the corresponding events in the log. Using this information, one can determine the processes associated with unauthorized events. By tracing the parent–child process relationships, one can often determine the vulnerabilities exploited and assess the damage caused by the attack. Then the user-process associations can be used to identify the user account(s) from which the violation of policy occurred. In the example where Robin's system was compromised, the logs would indicate that the mailer, and the web page, was modified.

The state-based approach analyzes the current state of the system to determine if the state is conforms to the security policy. A state includes the contents of files and the access rights of users. Consider how the state-based approach could protect the files stored on the web server machine, including Robin's web page, using a file integrity-checking tool.

Tripwire is an example of a file integrity-checking tool. Having seen the remote logon from Robin's machine to the web server (above), checking the web server to see if any files changed without authorization seems prudent. The file integrity-checking tool reports that Robin's web page was modified (presumably, without authorization). System configuration/vulnerability checkers such as COPS and SATAN fall into the class of state-based analysis tools; these vulnerability checkers analyze a system/ and report any known vulnerabilities.

8. RECOVERY

Recovery uses the assessment results to bring the system to a secure state. This includes terminating on-going attacks to stop further damages, replacing corrupted files with uncorrupted copies, removing system vulnerabilities to protect the system against future attacks, taking appropriate actions such as notifying affected parties or aborting planned actions, and restarting system services that have been made unavailable.

Suppose the Adversary's intrusion affecting Robin's web page is detected as it is occurring. A possible response is to terminate the attack process or the associated network connection to prevent the Adversary from committing further damage, and to take the altered web page offline to prevent other people from accessing it. The vulnerability that allows Robin's machine to be compromised must be fixed. For example, if the Adversary broke into the system by guessing Robin's password, that password needs to be changed. Following the assessment, the modified mailer and the corrupted web page will be replaced by a backup that has been stored in a secure place.

A common technique used for recovery is rollback—restoring a prior (secure) state of the system. In the above example, one can bring the web page to an earlier intact state because a backup version of it is available.

In practice, backup is usually done periodically. One can do a complete backup of all the files in the system, or one can do a selective backup in which only recently modified files or critical files are backed up. Different levels of backup may be combined—one may perform complete system backup once a week, and perform selective backups once a day. With respect to complete backup, performing a selective backup is cheaper but does not cover all files. To recover, one may need to use the last complete backup together with all successor selective backups to reconstruct the state of the system. The frequency of the backup is important because during rollback, all the changes made since the last backup may be lost. For programs that do not change over time, we may not need to make backups for them if we have the programs on distribution disks. For example, if the original mailer is secure, one can reload a clean copy of the mailer from the distribution disks. Note that this rollback technique is useful even if complete damage assessment is not possible.

Reconfiguration is a recovery technique in which the system is modified to bring it to a secure state. Reconfiguration is appropriate when one cannot roll back to a secure state, possibly because backups have not been done recently or the system has been in an insecure state since its inception. Changing Robin's cracked password to a more secure password is an example of reconfiguration.

Many vendors aid recovery by distributing "patches" or fixes for software with known vulnerabilities; this can be pre-emptive, because often system administrators receive patches before the vulnerability has been exploited. But sometimes nonsecure software or services cannot be fixed (because the flaw is one of interaction between the software and another component, and a fix would require modification of the operating system) or a fix may not be available. In those cases, administrators may simply disable the offending software or service. As an example, if the password of an account has been compromised, the owner of the account must change the password before the account can be used again. Freezing the account before the password change can prevent future attacks through that compromised account.

9. CONCLUSION: WHAT OF THE FUTURE

The state of security on the Internet is poor. The Internet infrastructure provides no supporting security services; the security of individual systems on the Internet varies wildly. But as the need for security on the Internet increases, new mechanisms and protocols are being developed and deployed. How will they improve the situation? Will attacks become more difficult to execute, easier to prevent, and simpler to detect and trace?

System security is a function of the organization that controls the system, so whether those systems will become more secure depends entirely upon the vendors and the organizations that buy their systems. As for the supporting services of the Internet, some improvement is taking place. Many new protocols have security services relevant to the service they provide, and application-level servers often try to use these services. Ipv6, the successor to the current underlying Internet protocol (IPv4), has security enhancements for confidentiality, authentication, and integrity built in. So there is hope.

In terms of preventing intrusions, the science of intrusion detection is one of the most active branches of computer security. The field of incident response is beginning to mature, as system administrators and researchers study ways to trace, analyze, prevent, and ameliorate intrusions. However, due to the complexity of systems, and thus the possible vulnerabilities in the design, implementation, configuration, and interaction of the hardware, software, and network protocols, vulnerabilities will continue to exist and be introduced anew.

The claim that a secure computer can be built and administered is at best questionable; perfect security in an imperfect world is unlikely because such a system can assume nothing, and trust nothing. And ultimately, the security of the Internet lies in the issue of trust. Policies require trust of low-level components (such as hardware and software) if nothing else; and people must trust the computer to function correctly, because it is impractical to verify every line of code, and all possible system states. Ultimately, people will determine what to trust; and so security is a nontechnical, people problem, deriving its effectiveness from the specifiers', designers', implementers', and configurers' understanding of what may be trusted, and how.

ACKNOWLEDGEMENTS

The authors are at the Computer Security Laboratory of the Department of Computer Science at the University of California, Davis. Matt Bishop, a member of the IEEE, is on the faculty of the Department and does research in computer and network security. Steven Cheung is a Ph.D. student in the Department and his doctoral research concerns network intrusion detection. Jeremy Frank recently received his Ph.D. from the Department and is currently working at NASA Ames Research Center as an employee of Caelum Research Corp; his current interests include combinatorial search techniques and problems in areas ranging from satellite networks to program synthesis. James Hoagland is a Ph.D. student in the department and does research in computer and network security, focusing on intrusion detection and policy language. Steven Samorodin is a graduate student in the Department and is doing research in the area of computer security. Chris Wee is a postdoctoral researcher in the Department and works in the areas of audit and intrusion detection.

SUGGESTED READING

Anderson, J. P. (1980). *Computer Security Threat Monitoring and Surveillance*. Fort Washington, PA: James P. Anderson Co., Apr.

This seminal paper introduced the idea of using audit logs to detect security problems,. It is the basis for intrusion detection and auditing.

Blaze, M., Feigenbaum, J., Lacy, J. (1996). Decentralized trust management. Proceedings of the IEEE Conference on Security and Privacy. pp. 164–173 May.

This paper discusses trust in a distributed environment, and illustrates the complexities of managing trust. As trust is the basis for all security, the theme of this paper speaks to a fundamental issue.

Cheswick, B. (1992). An Evening with Berferd in which a cracker is lured, endured, and studied. Proceedings of the Winter 1992 USENIX Conference. pp. 163–174, Jan.

This paper presents an encounter with an attacker who attempted to penetrate a Bell Labs system and was spotted. Rather than block the attack, the authors decided to allow the attacker access to a controlled environment to see what he or she would do.

Dean, D., Felten, E., Wallach, D. (1996). Java Security: from HotJava to netscape and beyond. Proceedings of the 1996 IEEE Symposium on Security and Privacy. pp. 190–200, May.

This paper discusses several security problems in various Java implementations and in the design itself. It illustrates the complexities of downloadable executable code, a subject far more complex than it would seem at first blush.

Denning, D. (1992). A new paradigm for trusted systems. Proceedings of the Fifteenth National Computer Security Conference. pp. 784–791, Oct.

Denning discusses trust and how it can be misplaced, and what the effect of misplacing it can be.

Staniford-Chen, S., Heberlein, L.T. (1995). Holding intruders accountable on the Internet. Proceedings of the 1995 IEEE Symposium on Security and Privacy. pp. 39–49, May.

Tracing a connection through the Internet is very complex, and often impossible. This paper proposes a statistical technique to correlate two connections to see if they belong to the same session.

Stoll, C. (1990). *The Cuckoo's Egg: Tacking a Spy Through the Maze of Computer Espionage*. New York: Pocket Books.

Stoll detected an attack on the Lawrence Berkeley Laboratory's computer when he spotted a 75 cent accounting error. The resulting investigation spanned two continents and led to the breaking up of an espionage ring. This book tells the story.

7

Computer Security: The Good, the Bad, and the Ugly

CATHERINE MEADOWS
Naval Research Laboratory,
Washington, DC, USA

ABSTRACT

In this paper we discuss and characterize different types of solutions to computer security problems in terms of bad (theoretically sound, but expensive and impractical), ugly (practical, but messy and of doubtful assurance), and good (theoretically sound and practical). We also attempt to characterize the different approaches and problems in computer security that would lead to these different types of solutions.

1. INTRODUCTION

Security of a system can be loosely defined as the assurance of a correct operation in face of a hostile attack. The fact that the threat is a hostile, intelligent, attacker has a number of ramifications:

1. How the system performed historically may have little bearing on how it will perform in the future. Although a system may have suffered few security problems in the past, this does not mean that this will always be the case. New applications of the system may make it a more attractive target for break-ins. New attacks and security holes may be found.
2. It is not always possible to predict what the security needs of a system will be, since it is not always possible to predict how the system will be used. For example, the Internet was originally intended to be a system by which a relatively small number of researchers could share resources, but has evolved into a communications system used by millions of people.
3. The benefits of security are usually invisible. Security controls cause bad things not to happen. Thus, users are often not willing to pay for security until they see evidence that these bad things will happen without it.
4. Finally, it is difficult to retrofit a system for security. Security controls must be safe against intelligent attack. This makes designing security controls "on top of" insecure systems a losing proposition, since the attacker can subvert the controls by attacking the underlying insecure system itself.

It is readily apparent that the implication of the first three statements is that security should be introduced into a system after it has been fielded and the security threats are understood. On the other hand, the implication of the fourth statement is that security should be designed into a system from the start. As can be imagined, these contradictory requirements can cause problems. Research has traditionally focused on designing security into a system from the start, since that is the best way of designing a system with insubvertible security controls. But user demand has generally led to security being added on after the fact. Indeed, since the security needs of a system depend at least in part upon the way in which the system is used, and since the way a system is used is often not easy to predict, it is often not possible to predict completely the security needs of a system until it has been in use for a while. This situation has led to elegant research results that are often beside the point, and fielded systems that are messy and of doubtful assurance. In other words, we generally have to make do with solutions that are bad in the sense that, although they may be theoretically sound, the do not address user needs, or ugly in the sense that, although they may be usable, they are messy and provide only limited protection. However, in certain cases it is possible to do good. In this paper, I will illustrate my point by giving examples of all three.

2. BAD

A well-known example of a bad solution, that is, one that is sound but expensive, is the Trusted Computer Systems Evaluation Criteria (Department of Defence Computer Security Center, 1983), commonly known as the Orange Book, issued by the Department of Defense in the early 80s. This provided sets of criteria that an operating system should satisfy to be certified for use at different levels of security. The Orange Book was mainly intended to provide criteria for secure operating systems used by the government, but there was also a category intended for the private sector. The key idea behind the Orange Book was the use of a security kernel that received requests for access and granted or denied it according to the security policy. The Orange Book also specified what kinds of policies could be enforced. There were two types: mandatory access control (MAC), which was used protect classified data, and discretionary access control (DAC), which was used to protect unclassified data. The Orange Book divided its criteria into a number of classes; the higher the class, the more security functionality and assurance were required.

This approach provided a solution that, although it had some shortcomings, was largely sound. However, this solution was achieved at a cost. Although the performance impact of the security overhead was not necessarily that great, the impact on development cost from the evaluation overhead could be huge, especially for classes that required a high degree of assurance. Moreover, in order to satisfy the Orange Book requirements, the operating systems often had to break existing interfaces, thus raising the cost of porting software. Finally, the MAC and DAC policies mandated by the Orange Book were too rigid for many applications. For these and other reasons, Orange Book compliant operating systems, especially at the higher levels of assurance, are not seeing nearly as much use as was originally planned.

3. UGLY

A well-known example of an ugly solution is the firewall (Cheswick and Bellowin, 1994). A firewall is a system that sits between a host network and the rest of the world, and decides what traffic will be allowed into the network. It may only allow traffic from particular addresses, or only allow certain types of applications to come through; for example, it might allow e-mail to get through but not executable code. This monitoring, of course, is always imperfect. It is possible to spoof addresses if traffic is not authenticated, and it is not always easy to determine which application is which. Moreover, firewalls do not protect against certain types of denial of service attacks (Rosen, 1996).* Thus, the security offered by firewalls

*Recall that this paper was written in 1996. But even today, fiirewalls can only mitigate against denial of service vulnerabilities caused by bad protocol design.

is coarse and imperfect; harmful traffic may be let in, and harmless traffic is often denied entrance.

However, firewalls have the advantage that they have little effect on the use of the host network within the security perimeter. Thus, it is possible to install a firewall with minimal cost to day-to-day operations, although it will put some limitations on the type of contact possible with the outside world. For this reason, firewalls have become almost ubiquitous, even though the security they offer is imperfect.

4. GOOD

Now we come to the "good" solutions. The first was largely developed at my own place of work, and was presented at this workshop; to lessen the chance of bias, I will give another one as well. The first example is the use of replicated data to provide capability to process classified data at different security levels. The basic idea is to have one system for each level. If the level for one system is greater than that of another, the data from the low system are replicated at the higher level. This allows the users of the high system to see low data without having to log in to the low system. However, the security risk is minimized because the low system does not receive data from the high one. In some versions, it is possible to have the high system send acknowledgments to the low system to increase reliability; in this case, it is possible to engineer tradeoffs between security and reliability requirements.

This approach was originally suggested for use in multilevel database management systems (Committee on Multilevel Data Management Security, 1983), and developed as such in the SINTRA multilevel secure database management system (Kang et al., 1994). It has also been used for more general data processing systems, such as Froscher et al. (1995).

This solution is a good one for a number of reasons. It is simple. The only thing that needs to be evaluated for security is the component that provides communication between the high and low systems, which can be as simple as a one-way optical link (Goldschlag, 1996) . It does not break interfaces. As a matter of fact, the most likely application of it so far has to hook up two or more systems that had previously been operating at system high at different levels, see for example, Kang et al. (1997).

My second example also involves multilevel security. This is the Starlight Interactive Link (Anderson et al., 1996), which provides a multilevel windowing capability by linking a workstation to classified and unclassified networks, and allowing the user to switch from one to another. The only trusted parts of the system are the Interactive Link itself, and a trusted display that lets the user know the current security level of the Link. Again, the solution is simple, does not break interfaces, and does not rely upon an insecure infrastructure.

What is it that we can learn from these different solutions? Perhaps it is most interesting to look at, not the solutions themselves, but the problems they have addressed. Orange Book systems were intended to replace insecure systems with secure ones. This meant that they broke existing interfaces that did not support security. Firewalls are intended to be a patch to existing insecure systems. Thus, instead of breaking interfaces, they attempt to deal with them as they exist, which means that they provide security that is either too strong (ruling out harmless accesses) or too weak (admitting accesses that could be harmful). The two approaches appear very different, but they both address the same problem: dealing with existing insecure systems, either by replacing or patching them.

On the other hand, the intended uses of the replicated data technique and the Starlight Interactive Link have been to supply secure communication to system-high systems that were previously isolated or connected only by sneaker net. In other words, instead of replacing or fixing an insecure system, they are adding a secure capability that did not exist before. The lesson from this appears to be that it is possible to obtain a good solution when we attempt to provide a secure connection where none existed before, but much more difficult when we attempt to replace insecure connections with secure ones. In that case, one must either deal with or replace the existing infrastructure.

This conclusion seems to be in line with the observation made earlier in this paper that security is best added on early in a system's lifetime. Since we have also noted that security issues are often not addressed until after the fact, our conclusion may seem to be somewhat pessimistic. But it is not quite as pessimistic as it seems. It points out that, even if a system has already been use for some time, it may be possible to add security without resorting to bad or ugly mechanisms if we provide a new capability that is secure instead of attempting to modify existing ones. Perhaps this observation can be used to guide other attempts to bring security to existing systems.

REFERENCES

Anderson, M., Nort, C., Griffin, J., Milner, R., Yesberg, J., Yiu, K. (1996). Starlight: interactive link. In: Proceedings of the Twelfth Annual Computer Security Applications Conference. IEEE Computer Society Press, Dec., pp. 55–63.

Cheswick, W. R., Bellovin, S. M. (1994). *Firewalls and Internet Security*. Reading, MA: Addison-Wesley.

Committee on Multilevel Data Management Security, Air Force Studies Board, Commission on Engineering and Technical Systems, National Research Council. (1983). *Multilevel Data Management Security*. Washington, DC: National Academy Press.

Department of Defense Computer Security Center (1983). Department of Defense Trusted System Evaluation Criteria. Aug. CSC-STD-001-83.

Froscher, J. N., Goldschlag, D. M., Kang, M. H., Landwehr, C. E., Moore, A. P., Moskowitz, I. S., Payne, C. N. (1995). Improving inter-enclave information flow for a secure strike planning application. In: Proceedings of the 11th Annual Computer Security Applications Conference. Dec.

Goldschlag, D. M. (1996). Several secure store and forward devices. In: Proceedings of the Third ACM Conference on Computer and Communication Security. ACM, Mar.

Kang, M. H., Froscher, J. N., McDermott, J., Costich, O., Peyton, R. (1994). Achieving database security through data replication:

the SINTRA prototype. In: Proceedings of the 17th National Computer Security Conference. Baltimore, MD, Sept., pp. 77–87.

Kang, M. H., Froscher, J. N., Moskowitz, I. S. (1997). A framework for MLS interoperability. In: Proceedings of HASE 97.

Rosen, A. (1996). Understanding and defending against SYN attacks. In: Proceedings of the DIMACS Workshop on Network Threats.

8

Architecture and Applications for a Distributed Embedded Firewall

CHARLES PAYNE and TOM MARKHAM
Secure Computing Corporation,
Roseville, CA, USA

ABSTRACT

The distributed firewall is an important new line of network defense. It provides fine-grained access control to augment the protections afforded by the traditional perimeter firewall. To be effective, though, a distributed firewall must satisfy two critical requirements. First, it must embrace a protection model that acknowledges that everything behind the firewall may not be trustworthy. The malicious insider with unobstructed access the network can still mount limited attacks. Second, the firewall must be tamper-resistant. Any firewall that executes on the same untrusted operating system that it is charged to protect begs the question: who is protecting whom? This paper presents a new distributed, embedded firewall that satisfies both requirements. The firewall filters Internet Protocol traffic to and from the host. The firewall is tamper-resistant because it is independent of the host's operating system. It is implemented on the host's network interface card and managed by a protected, central policy server located elsewhere on the network. This paper describes the firewall's architecture and associated assurance claims and discusses unique applications for it.

1. INTRODUCTION

Traditional perimeter firewalls are a critical component of network defense, but they should not be considered the only line of defense. First, their protection is too coarse. This leaves the firewall helpless against the malicious insider, who operates freely within the firewall's security perimeter. Second, it is costly to extend their protections to mobile users, because the firewall's security perimeter is determined somewhat artificially by the firewall's location in the network topology. For effective network defense, we must augment perimeter firewalls with more fine-grained access controls.

Bellovin (1999) argued that a distributed firewall provides the fine-grained protection that is needed. In this solution, a firewall is placed at each host in the network, and all firewalls are managed as a single entity. That is, centralized management is coupled with distributed enforcement. Distributed firewalls contain the malicious insider because the security perimeter is drawn around each host. Because the perimeter is no longer defined by network topology, the distributed firewall is an ideal solution for mobile users, telecommuters and business-to-business extranets. Also, since distributed firewall policy is expressed in terms of network endpoints, changes to network topology have little if any impact on policy management. Ioannidis et al. (2000) described a prototype distributed firewall for Open BSD hosts.

The distributed firewall falls short, however, if it assumes that all users on the local host are trustworthy. If these users are trusted to access the network freely, limited attacks such as network sniffing, host address spoofing, and denial of service are still possible. To effectively contain the malicious insider, distributed firewalls must embrace a stronger protection model that acknowledges that users on the host may not be trustworthy. In other words, in addition to protecting the host from a malicious network, the firewall must protect the network from a malicious host. This requirement becomes more significant when we recognize that many insider attacks are not mounted intentionally. Worms like Code Red and NIMDA, for example, can turn loyal users into unwitting insiders (Markham and Payne, 2001).

The distributed firewall also falls short if it executes on an untrusted operating system. So-called personal firewalls suffer this fate. These software-based solutions fail to satisfy a cornerstone requirement for firewalls: tamper resistance. Personal firewalls are relatively easy to disable via a network-based attack (White, 2001). Like the emperor's dressmaker, they can leave the host—and by consequence the network—a bit exposed.

This paper describes a new distributed embedded firewall called EFW that embraces the stronger protection model and that is tamper-resistant. EFW filters Internet Protocol (IP) traffic to and from the host. EFW is tamper-resistant because it is independent of the host's operating system. Instead, the firewall is implemented on the host's network interface card (NIC) and managed by a central, protected policy server elsewhere on the network. EFW is implemented on a commodity NIC and scales easily to thousands of hosts.

The paper focuses on EFW's architecture and how it can support interesting security applications. Section 2 defines EFW's security and non-security objectives. Sections 3 and 4 illustrate the component and management architectures, respectively, that result from these objectives.

Distributed firewalls like EFW offer new opportunities in security policy enforcement. Section 5 enumerates several novel applications that we have identified for EFW.

Finally, EFW's genesis occurred in DARPA-sponsored research from the late 1990s. Since then, we have investigated its use in many problem domains. Section 6 offers a glimpse at EFW's future directions.

The remainder of this section considers related work.

1.1. Related Work

While EFW's goals and objectives closely resemble those of Bellovin (1999), the two efforts proceeded independently and resulted in very different implementations (Bellovin's implementation is described in Ioannidis et al., 2000). Bellovin noted that "for more stringent protections, the policy enforcement can be incorporated into a tamper-resistant network card" (Bellovin, 1999, Section 7.5), but he chose to implement his distributed firewall with kernel extensions, a user level daemon and a new device driver. Besides offering a simpler development path for a prototype, this strategy enabled the firewall to handle application-level policies. EFW, on the other hand, focuses on IP packet filtering because of the limited processing power available on the NIC.

Nessett and Humenn (1998) proposed a novel multilayer firewall that can be managed centrally. Nessett's firewall includes all of the devices, such as perimeter firewalls, switches, and routers that currently perform filtering in the network. This work illustrates the pitfalls that can be encountered when firewall policy management is inextricably bound to network topology management. Bellovin (1999) and Markham and Payne (2001) advocate breaking this bond. The results of Nessett and Humenn also underscore the importance of creating multiple layers (e.g., distributed firewalls and perimeter firewalls) in an overall network defense strategy.

2. OBJECTIVES FOR EFW

We divide the objectives for EFW into two camps: security-related and non-security-related.

2.1. Security Objectives

Figure 1 illustrates an EFW NIC on a protected host, the EFW policy server (also protected by an EFW NIC), and the communication paths between them. To illustrate our high-level design strategy, we state the security objectives for EFW in the form recommended by Payne et al. (1993). Assertions that EFW must satisfy are expressed as claims. Following the statement of each claim are zero or more assumptions upon which the claim relies. Validated assumptions are represented as claims elsewhere in this section and are so referenced. Unvalidated assumptions represent potential vulnerabilities for EFW that must be validated by other means (procedural controls, physical security, and so forth).

The top-level claim is that EFW performs its function correctly.

Claim 1 EFW blocks unapproved IP traffic to and from the host, which assumes:

- EFW is configured properly, and
- EFW is non-bypassable (Claim 2).

The first assumption captures the importance of strength in policy, while the second assumption captures the importance of strength in mechanism. The first assumption

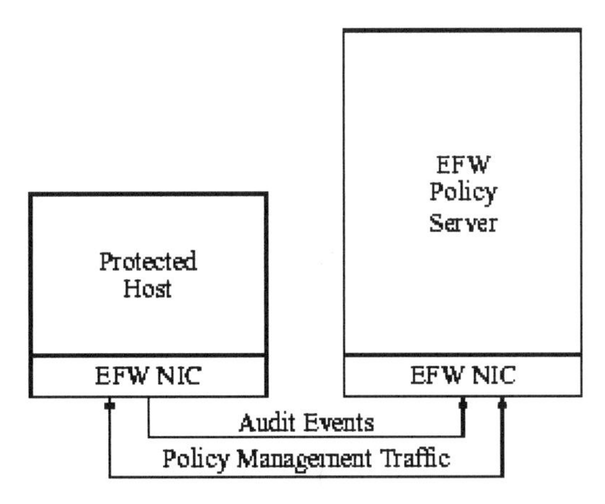

Figure 1 EFW NIC and policy server.

is validated on a case-by-case basis. Essentially we must ensure that the policy enforced by EFW is appropriate for the operational environment and its security threats. We will not consider this requirement further except to describe, in Section 4, the tools that EFW provides for policy management.

Claim 2 EFW is non-bypassable, which assumes:

- The host can communicate only through EFW-enabled NICs.
- EFW is tamper-resistant to host-based attack (Claim 3).
- EFW is tamper-resistant to network-based attack (Claim 4).

The first assumption is not trivial to achieve, and currently EFW offers no technical means to validate it. This means that we cannot, for example, stop the user from swapping out the EFW NIC for a non-EFW NIC. However, technical measures do exist in the EFW policy server to detect such activity. Fortunately, this potential vulnerability is temporary. Section 6 describes a technology that will prevent the host from accessing network resources unless it communicates through an EFW NIC.

Claim 3 EFW is tamper-resistant to host-based attacks, which assumes:

- The EFW NIC hardware is protected from direct manipulation.
- Only the EFW policy server can disable an EFW NIC (Claim 5).
- Only the EFW policy server can download new policy (Claim 6).
- Attackers cannot masquerade as the EFW policy server (Claim 8).

Claim 4 EFW is tamper-resistant to network-based attacks, which assumes:

- Only the EFW policy server can disable an EFW NIC (Claim 5).

- Only the EFW policy server can download new policy (Claim 6).
- Attackers cannot masquerade as the EFW policy server (Claim 8).

The first assumption in Claim 3 can be validated by restricting the hardware interfaces. Newer generations of the 3CR990 NICs take steps in that direction by combining more functions onto fewer chips. The remaining assumptions for Claim 3 also apply for Claim 4. That is not a coincidence. While we typically imagine the EFW NIC as being managed remotely, the EFW policy server can protect itself with an EFW NIC, which it will manage locally. The protection mechanisms implemented on the EFW NIC do not distinguish whether the policy server is local or remote, and the host enjoys no privileged access to the EFW NIC.

Similarly, the next two claims have identical supporting assumptions.

Claim 5 Only the EFW policy server can disable an EFW NIC, which assumes:

- The operation is available only by a command to the EFW NIC API.
- The command is accepted only from the EFW policy server (Claim 7).
- Only authorized users can access the EFW policy server.

Claim 6 Only the EFW policy server can download new policy to an EFW NIC, which assumes:

- The operation is available only by a command to the EFW NIC API.
- The command is accepted only from the EFW policy server (Claim 7).
- Only authorized users can access the EFW policy server.

The first supporting assumption for Claims 5 and 6 is validated by the EFW implementation. The third supporting assumption can be validated by procedural controls and physical security. The remaining assumption is validated by a combination of technology, procedural and physical security controls, as described below.

Claim 7 The command is accepted only from the EFW policy server, which assumes:

- All policy server/NIC communications is authenticated by 3DES (Claim 9).
- Only the EFW policy server and the EFW NIC possess the cryptographic key (Claim 10).

The remaining assumption from Claims 3 and 4 is validated by the same controls.

Claim 8 Attackers cannot masquerade as the EFW policy server, which assumes:

- All policy server/NIC communication is authenticated by 3DES (Claim 9).
- Only the EFW policy server and the EFW NIC possess the cryptographic key (Claim 10).

The next claim defines the technology controls.

Claim 9 All policy server/NIC communication is authenticated by 3DES, which assumes:

- The work factor to break 3DES is too high.

The last claim defines the procedural and physical security controls.

Claim 10 Only the EFW policy server and the EFW NIC possess the cryptographic key, which assumes:

- Only authorized users can access the EFW policy server.
- EFW crypto keys are protected from compromise.

2.2. Other Objectives

While most of the research behind EFW was funded by the US Department of Defense (DoD), the DoD relies increasingly on commercial-off-the-shelf solutions, so the needs of DoD and the commercial marketplace are not dissimilar. As a result, we also considered commercial viability and commercial acceptance throughout this effort. In addition to being secure, we determined that EFW needed to be cost-effective, scalable and friendly to manage.

2.2.1. Cost-effective

The constraints of implementing on an NIC prompted the motto: "fast, simple, and cheap". For performance, an NIC has a tight processing loop, and our solution had to fit within those bounds. An NIC also has limited memory, so complex processing is performed elsewhere (e.g., on the EFW policy server). Lastly, the 3CR990 an NIC is relatively inexpensive, and we did not want to alter that fact. These NICs are already widely deployed, so modifications to the existing hardware and its drivers were avoided at all costs. We confined our modifications to the NIC's firmware.

2.2.2. Scalable

To facilitate commercial acceptance, it must be possible for administrators to introduce EFW as little or as much as they like. Initially, some administrators may prefer to protect only a few critical servers; others may immediately deploy EFW to every client desktop along with a policy that enforces "good network hygiene". The differences between a large deployment (thousands of NICs) vs. a small one are minimized in the eyes of the administrator through the use of management abstractions (explained in Section 4). We also adopted a master/slave architecture between the policy server and its NICs.

2.2.3. Friendly to Manage

To make EFW friendly to manage, we created several administration tools, including a policy editor, an EFW device (NIC) manager, and an audit logger and event viewer. These tools rely on several abstractions to reduce their complexity. Our objective was to make EFW invisible to the end user and to incorporate familiar management paradigms for the administrator.

3. IMPLEMENTING EFW

The high-level architecture for EFW is illustrated in Figure 2. The protected host may be a client workstation, a server, or any other device that supports the NIC. The policy server should be installed on a dedicated host and protected by its own EFW NIC. The following sections describe the components on each platform in greater detail.

3.1. EFW Components on the Protected Host

Three components reside on each protected host: the EFW-enhanced NIC, the NIC's driver and runtime image, and a non-security-critical helper agent.

3.1.1. EFW-enhanced NIC

The most important component on the protected host is the NIC and its EFW-enhanced firmware. EFW is based on the 3Com 3CR990 family of NICs. We selected these NICs for several reasons. First, they have an on-board processor and memory, which allows the NIC to operate independently of the host operating system. Second, they contain an on-board cryptographic engine. This feature was included to support

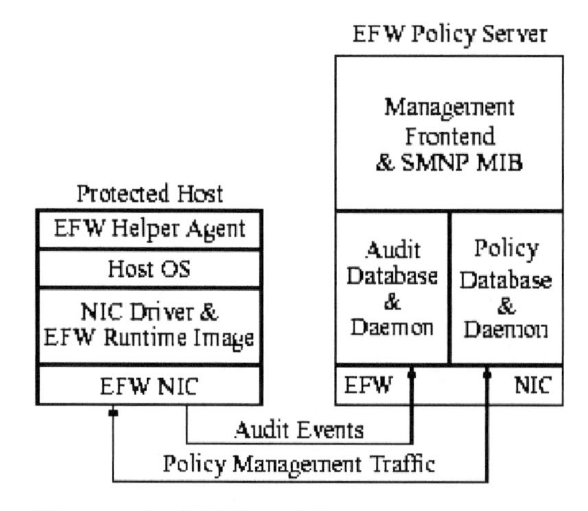

Figure 2 EFW architecture.

Windows 2000 IPSEC offloads, but EFW also leverages the crypto engine to provide secure communications with the policy server. Finally, these NICs are relatively inexpensive and widely available.

Flashed onto the NIC during the EFW install, the EFW-enhanced firmware contains the packet filtering engine and the management interface for the EFW policy server. The packet filter can accept or reject packets according to the standard parameters (source and destination address, source and destination port range, IP protocol, packet direction, etc.) as well as the value of the TCP SYN flag (used for connection initiation) and the presence of IP options. It can also accept or reject fragmented packets and non-IP packets. Each filter rule can be configured to generate an audit event. The management interface handles policy downloads from the policy server and transmits audit events to the audit server. It is also responsible for managing the secure channel with the policy server.

3.1.2. Driver and Runtime Image

The driver installed for an EFW NIC is the unmodified commercial driver. Like similar products, this NIC relies on its driver upon each host reboot to download its runtime image into the firmware. To ensure that a host remains protected, the EFW NIC stores enough information in non-volatile memory to verify the integrity of its runtime image. Once an NIC is configured for EFW, it cannot be disabled except by performing the appropriate action on the policy server. In other words, the EFW NIC will become inoperable if its runtime image fails the integrity check.

3.1.3. Helper Agent

The NIC must know its IP address in order to enforce policy. In a DHCP environment, it will need the host's assistance to determine that address. A small helper agent in user space performs this function. The helper agent also sends regular heartbeats to the policy server to help the policy server detect NICs that may not be functioning. Like all other communications with the policy server, the heartbeat is encrypted by the EFW NIC. If a malicious user were to replace the EFW NIC with a vanilla NIC, the heartbeats for that EFW NIC would effectively stop, raising the suspicions of the EFW administrator. The EFW NIC does not rely on the helper agent for continued operation and will continue to enforce policy even if the helper agent crashes or is removed.

3.2. EFW Policy Server Components

The EFW policy server is composed of three main components:

1. the management component, including the graphical user interface (GUI), the SNMP management information base (MIB), and the controller front-end,
2. the policy component, including the policy daemon and the policy database, and
3. the audit component, including the audit daemon and the audit database.

3.2.1. Management Component

The management component is described more fully in Section 4. Its main purpose is to provide the administrator with the tools to create, view, and distribute policies to each EFW NIC. It also includes an audit browser to review event logs. The MIB will support future network management applications.

3.2.2. Policy Component

The policy component takes the policies defined using the management component and compiles them into filter rules for each NIC. This component ensures that NICs enforce the policy to which they are assigned. When an NIC's host is rebooted, the NIC requests the current policy from the policy server. If the policy server does not respond, the NIC "falls back" to enforce a simpler policy. Currently the choices are: allow all traffic, allow all traffic but prevent network sniffing, or deny all traffic. If a policy is modified, the policy component automatically pushes the updated policy to the affected NICs. NICs that are off-line during the policy push receive the policy once they return on-line. The heartbeat generated by the host's helper agent informs the policy component of the policy that the NIC is enforcing.

3.2.3. Audit Component

The audit component receives audit events from each NIC and stores them in a database for browsing and searching by the management component. Audit logs can also be exported to third-party tools for additional analysis. As the arrows in Figure 2 imply, policy updates from the policy component to the NIC are acknowledged by the NIC; however, the audit component does not acknowledge audit events generated by the NIC.

4. MANAGING EFW

EFW provides many useful abstractions to help the administrator define and manage policies. This section describes those abstractions and discusses the challenges and opportunities of managing distributed firewalls.

4.1. Abstractions

EFW divides protected hosts into policy domains. A policy server can manage only one policy domain, although there may be multiple policy servers for each domain. A policy domain might encompass an entire organization or perhaps just one or two divisions within that organization.

Within each policy domain, NICs are grouped by function into device sets. There might be one device set for managers, another for the finance staff, etc. Device sets reduce complexity by grouping together the NICs that are likely to be assigned the same policy. So while there might be thousands of NICs in a policy domain, there may only be a dozen or so device sets. Each device set is assigned a single policy, although a particular policy may be assigned to multiple device sets.

Policies are composed of policy attributes and rules. Policy attributes represent facts that apply across all rules of the policy. For example, "the host is not allowed to spoof its IP address", or "fragmented packets are not permitted". EFW rules are similar to the packet filter rules found on other firewalls. For convenience, rules can be grouped into rule sets. If any rule in the rule set is modified, the changes propagate to all policies that include the rule set.

EFW also supports audit and test mode. Audit can be set for an entire policy or just for an individual rule. Test mode works with audit to help the administrator understand the effects of a policy or a single rule before it is actually enforced. For example, a rule that is in test mode will generate audit events each time a packet matches it, but the action associated with that rule (allow or deny) will be ignored.

Audit is also very useful for "discovering" policy. For example, to identify the network services that a particular host requires to boot up and log on users to the net, we can push an "allow but audit" policy to its EFW NIC. Then we reboot the host and watch the audit logs. A network monitor would perform a similar function.

4.2. Challenges

EFW is not immune to the policy management challenges that face other packet filters, and the limited resources of the NIC make overcoming these challenges even more difficult. For example, port mapping protocols, i.e., protocols that start on a well-known port then negotiate a higher, random port to complete the session, require the firewall to maintain some state about the session. Protocols that use a well-known control port and a random data port (e.g., FTP and some streaming media protocols) are similarly challenging. We are examining alternatives for solving this problem for EFW. Fortunately, the challenge exists only if we need to specifically allow these protocols while denying everything else. If we want to deny these protocols, we can deny the connection to their well-known ports.

4.3. Opportunities

Managing policy for a distributed firewall like EFW is not simply a matter of moving the perimeter firewall's policy to each endpoint EFW device. Consider the following policy:

> Allow HTTP requests from a specific client to a specific web server.

On a traditional firewall, this policy might be stated as a single rule (see Table 1), where a rule is stated in the form (action, protocol, port, source, destination). We assume that traffic is permitted in both directions.

Placing this rule on both the EFW for the client and the EFW for the web server would be redundant, and if it was the only rule enforced by either EFW, it would probably be overly restrictive. More likely, the policy writer would choose to distribute the policy between the two devices, such as illustrated in Table 2. However, the same effect could be achieved by distributing the policy as in Table 3. In both tables, the web server is restricted to processing HTTP requests. However, in Table 2, the client may make other requests, while in Table 3, the client is restricted to HTTP requests.

Tables 2 and 3 express different policies from each other and from Table 1; however, the differences are only evident when we express the policies for EFW. The traditional firewall policy did not specify the behavior of the client and the web server beyond HTTP requests flowing through the perimeter firewall. For a particular site, these distinctions may be important, and EFW helps us to make them. EFW lets the administrator state policies far more precisely.

5. EFW APPLICATIONS

While EFW can certainly handle applications conceived for traditional, packet-filtering firewalls, its real power lies in

Table 1 Traditional Firewall

(Allow, TCP, 80, client, web server)

Table 2 EFW—Option 1

for host client
(Allow, *,*, client, *)
for host web server
(Allow, TCP, 80, *, web server)

applications that are either not possible or not feasible using traditional firewalls. This section describes several useful applications that we have encountered. Each application forms a building block that can be used to construct even more interesting applications.

5.1. No Sniffing, No Spoofing

One of the most significant applications for EFW is its ability to enforce good network hygiene. In general, a host should not be able to sniff other network traffic or spoof its IP address to other hosts. Many network attacks rely on one or both behaviors. Distributed denial of service attacks, for example, direct zombies to flood the victim host with spoofed service requests. EFW can prevent the NIC's untrusted driver from placing the NIC in promiscuous mode, and it can also prevent any packet from leaving the host that is not tagged with a valid IP address for that host.

5.2. Lock Down the Host

One of the biggest problems for IT personnel is keeping up with the security patches that must be installed. Often these patches are for services that are installed by default when the

Table 3 EFW—Option 2

for host client
(Allow, TCP, 80, client, *)
for host web server
(Allow, TCP, 80, *, web server)

operating system is installed. Sometimes the services are network services that the user should not be invoking anyway. Rather than manually reconfiguring each host to disable the service, EFW can prevent the service from being available. Another problem is users who configure their hosts in violation of the organization's security policy. For example, most network administrators prefer that users share files using an IT-maintained resource. However, a user can easily configure the typical PC to share files from the local disk. EFW can prevent this behavior by preventing file access requests from reaching the host.

5.3. Servers are Not Clients

Dedicated servers should not perform certain functions normally reserved for clients, such as sending email, making web requests, and so on. The NIMDA worm, for example, relies on this behavior to propagate. For TCP-based services, the easiest defense is to prevent the host from initiating TCP connections to other hosts. EFW can prevent unauthorized, outgoing TCP connection initiation requests from ever reaching the network.

5.4. Clients are Not Servers

Similarly, client hosts should not respond to service requests from other hosts. For TCP-based services, EFW can prevent unauthorized, incoming TCP connection initiation requests from ever reaching the host.

5.5. Stay in Your Own Backyard

With the exceptions of web and certain related traffic (FTP, streaming media, etc.), client hosts obtain most network services from dedicated LAN (local area network) servers. For example, it is typically not necessary for a user to access any DNS or SMTP server other than the one assigned by IT. To accomplish this goal, EFW can allow limited "external" requests, then restrict all other traffic to the local subnet.

5.6. Don't Talk to Strangers

If a particular network service, e.g., DNS, should be provided only by a specific server or group of servers, then EFW can restrict access to that service on only that server or group of servers.

5.7. Emergency Rule Set

This very useful application utilizes the rule set feature of the EFW policy server. Rule sets are included in policies by reference, not by value, so a change in the rule set propagates to all affected policies. Using this capability, an administrator can define an emergency rule set to be included in all policies. If a network attack is detected that requires a particular service port, the administrator can add a rule to the emergency rule set denying that port. With a click of a single button, the administrator can distribute this new restriction to all EFW NICs in the EFW policy domain.

5.8. Shared Server

Business-to-business communications require an infrastructure for sharing information. Extranets are the common solution, but extranets are expensive to implement. EFW enables a lightweight, cheaper alternative: the shared server. A single host with two EFW NICs is placed where it is accessible by both organizations. The organization that hosts the server controls both EFW NICs. We assume that both organizations may have administrator privileges to the server. The EFW NIC that is attached to the controlling organization's LAN prevents the shared server from initiating unauthorized communication on the LAN and sniffing traffic on the LAN. The EFW NIC that is attached to the Internet allows only protected communications with the business partner. The business partner can enter and access the shared server, but it is unable to exit out the "other side".

6. FUTURE WORK

As we gain more experience with EFW, we envision applications will require features not yet present in the architecture. For example, through our DARPA-sponsored research programs, we are currently investigating tie-ins to intrusion detection and response systems and using the EFW NICs to provide load sharing within server clusters.

Another important area of investigation is a technology we call virtual private groups (VPG) (Carney et al., submitted). Like a virtual private network (VPN), a VPG establishes a community of interest that is not restricted by network topology. Unlike the VPN, however, which establishes only pair wise relationships between the communicating entities, the VPG establishes group-wide relationships. The VPG architecture significantly simplifies key management for hosts within the group and makes management of secure group communications practical. When it is available, VPG technology will enable organizations to quickly set up, use, and then tear down secure group communications for wireless LANs and collaboration tools.

The VPG technology will also be an important catalyst for ensuring that network communication occurs only through EFW NICs (see the first assumption under Claim 2 in Section 2.1). If all hosts belong to one or more VPGs, and if network services are available only to members of the appropriate VPG, then only EFW NICs will be able to access network services. The attacker who attempts to access the network with an unsecured NIC will be completely thwarted.

7. SUMMARY

We have described a distributed, embedded firewall called EFW that is implemented on the host's network interface card. In addition, we have discussed several useful and unique applications for EFW. EFW can be used to lock down critical assets, such as corporate web servers, databases, and administrative workstations, and it can be used to lock down critical services, such as DHCP, DNS, and so forth. It lets the administrator easily control unnecessary capabilities on the network. Finally, EFW demonstrates that finer-grained network access control is possible and practical. Together with the perimeter firewall, it forms a strong line of network defense.

ACKNOWLEDGMENTS

The authors are grateful for the financial support of the US Defense Advanced Research Projects Agency. This paper reflects work performed under the Releasable Data Products Framework program (contract no. F30602-99-C-0125, administered by the Air Force Research Laboratory) and the Autonomic Distributed Firewall program (contract no. N66001-00-C-8031, administered by the Space and Naval Warfare Systems Center). The authors also wish to thank the anonymous reviewers for their helpful and insightful comments.

REFERENCES

Bellovin, S. M. (1999). Distributed firewalls.;login:, pp. 37–39, Nov.

Carney, M., Hanzlik, B., Markham, T. (2002). Virtual private groups. 3rd Annual IEEE Information Assurance Workshop, June 17–19, 2002.

Ioannidis, S., Keromytis, A. D., Bellovin, S. M., Smith, J. M. (2000). Implementing a distributed firewall. In: Seventh ACM Conference on Computer and Communications Security, Athens, Greece, Nov. ACM.

Markham, T., Payne, C. (2001). Security at the network edge: a distributed firewall architecture. In: DISCEX II, Anaheim, CA, June. DARPA, IEEE.

Nessett, D., Humenn, P. (1998). The multilayer firewall. In: Network and Distributed System Security Symposium, March.

Payne, C. N., Froscher, J. N., Landwehr, C. E. (1993). Toward a comprehensive INFOSEC certification methodology. In:

Proceedings of the 16th National Computer Security Conference, Baltimore, MD, Sep. NIST/NSA, pp. 165–172.

White, A. (2001). New trojan disables firewall defences. *Network News*, May (http: www.vunet.cou/NEWS/1125025).

9

Using Operating System Wrappers to Increase the Resiliency of Commercial Firewalls

JEREMY EPSTEIN and LINDA THOMAS
Product Security & Performance
webMethods, Inc., Fairfax, VA, USA

ERIC MONTEITH
McAfee Labs,
Herdon, VA, USA

ABSTRACT*

Operating system wrappers technology provides a means for providing fine-grained controls on the operation of applications software. Application proxy firewalls can gain from this technology by wrapping the proxies, thus preventing bugs (or

*The work described in this paper was performed while all three authors were associated with NAI Labs. This paper was published in Proceedings of the 16th Annual Computer Security Applications Conference, December 11–15, 2000, New Orleans, LA.

malicious software) in the proxy from subverting the intent of the firewall. This paper describes several experiments we performed with wrappers and firewalls, using several different firewalls and types of wrappers.

1. INTRODUCTION

Access controls in operating systems are usually at a coarse level and frequently do not cover all types of resources in the system. For example, UNIX systems control access to files, but the only controls on sockets limit nonroot processes from binding low numbered sockets. Operating system wrapper technologies (henceforth "wrappers"), including those described in Jones (1993), Fraser (1999), Balzer (1999), among others, allow specifying the behavior of application processes to an arbitrary level of granularity.[†]

While wrapper technology is aimed at constraining the behavior of applications on end systems (especially clients, and possibly also servers), it is also applicable to security devices such as firewalls. As part of the DARPA Information Assurance program, we have performed a series of experiments using different types of wrappers to constrain the behavior of several different firewall products. This paper describes the results of those experiments, and points to directions for future research.

The remainder of this paper is organized as follows. Section 2 describes our motivation for developing firewall wrappers. While this paper assumes a basic understanding of wrapper technology, Section 3 provides a synopsis of what wrappers are and how they work, and describes some of the differences between the wrappers technology developed by NAI Labs (Fraser, 1999), and the wrappers technology developed by the Information Sciences Institute (ISI) (Balzer,

[†]The term "wrappers" is overloaded in the security field. In this paper, it means functions that intercept system calls and perform mediation. This is different from TCP Wrappers (Venema, 1992) which are a program between *inetd* and the service provider daemons, but do not attempt to intercept system calls.

1999). Section 4 describes how we wrapped the Gauntlet Internet Firewall (for which we had design information and source code available) using the NAI Labs wrappers. Section 5 describes our experiences in using the NAI Labs wrappers to wrap firewalls for which we had no source code or design information. Section 6 describes how we wrapped the Gauntlet Internet Firewall using the ISI wrappers, and as such is a parallel to Section 4. Section 7 gives our current status and availability of our prototypes. Section 8 concludes the paper.

2. MOTIVATION

Application level firewall proxies are fragile, and are growing ever more complex. Customers demand increasing functionality, including the ability to perform tasks such as virus scanning, limits on addresses visited (e.g., to prevent access to pornographic web sites), and detailed scanning of protocols to prevent outsiders from exploiting vulnerabilities in host systems. As the proxies become increasingly complex, the likelihood of flaws that allow security breaches increases. For example, it is likely that there are opportunities in most firewall proxies for buffer over-run attacks.

As the number of protocols increases, proxies are increasingly written by people without sufficient training in writing safe software. End users want to write their own proxies, since they can do it more rapidly than waiting for a firewall vendor to include a suitable proxy in the product. While both vendors and end users make reasonable efforts to ensure that proxies are not being written by hostile developers (who might insert backdoors or other malicious software), it is likely that such capabilities have been inserted in at least some proxies. Finally, when this effort was begun, there was significant concern about the then-impending Y2K crisis, and as such there was concern in DoD and elsewhere that backdoors were being inserted as a byproduct of Y2K remediation, including in security products such as firewalls.

Since a single faulty proxy can endanger an entire firewall (and the network behind it), it is important to constrain the damage done by an errant proxy. The historical approach to such threats would be to use good software engineering techniques (including code inspection), personnel security (such as clearances), and improved testing. However, these are not realistic in today's "Internet time" commercial products environment. Additionally, we want to constrain proxies for which we may not have source available, since that allows us to use a variety of different products.[‡]

Our goal, then, was to provide a method for constraining proxies without requiring source code. The constraints should be simple enough (i.e., much simpler than the proxy itself), so that the wrapper can reasonably be subjected to a detailed correctness analysis. The wrappers technology allows us accomplish these goals.

Some argue that it is more effective to harden the operating system used in the firewall than to add a layer of protection such as wrappers. In a certain sense, wrappers are hardening the operating system: they allow for a more granular control of capabilities, but in a general purpose way that can serve not only firewalls but also other types of computer systems.

3. WRAPPERS SYNOPSIS

The idea of wrappers technologies is to provide for relatively small specifications of the allowed behavior of software. The NAI Labs wrapper technology is currently implemented for UNIX and NT; the ISI wrappers technology is currently available for NT only.

The premise of all wrappers technology is that the application being wrapped should be unaware of the wrapping, and should not need any modifications to be wrapped. Applications may become indirectly aware of wrappers because operations that succeed on an unwrapped system

[‡]While some object to the very concept of running proxies without source code available, it is the essence of today's commercial firewall market. Therefore, it is important to address, even if it is not the ideal situation.

may fail on a wrapped system, but this is an unavoidable side effect.

In some cases, it is possible to determine a priori what system calls will be used (e.g., by statically examining the object file to determine the interfaces invoked through a DLL or a shared library). In the more general case, though, some system call interfaces may have been statically linked, in which case such external references will not appear, or system calls may be called directly without going through APIs. Thus, it is important to catch the system calls where they enter the operating system, and not at an API level.

The UNIX version of the NAI Labs wrapper technology uses a kernel loadable module to intercept all system calls as they are made, and pass them to a wrapper based on criteria specified when the wrapper is loaded. We will refer to the former as the wrapper enforcement layer (WEL) and the latter as the wrapper specification (WS). A system can have many different wrapper specifications loaded, with different activation criteria. For example, a WS might apply to all programs run by a specific user, or to all instances of programs run from a particular directory. Figure 1 shows the operation of the wrappers technology. Because the interception occurs inside the kernel, the NAI Labs wrappers cannot be bypassed by malicious code.

The NT version of the NAI Labs wrapper technology and the ISI wrappers technology rely on intercepting system calls in user space, before they are passed to the NT kernel. Both

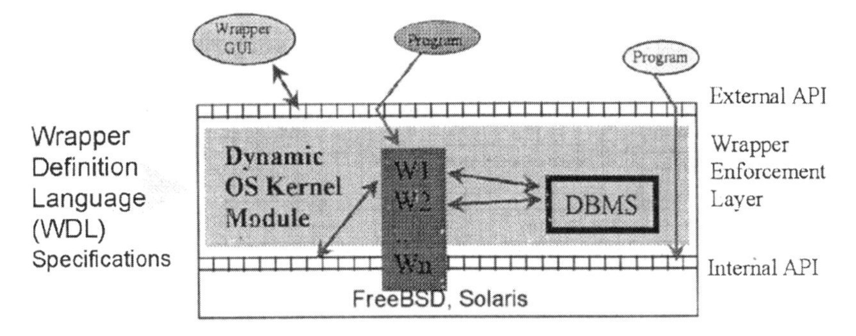

Figure 1 NAI Labs wrapper technology.

technologies strive to make it as difficult as possible to bypass the wrapper, but since they operate in user space, they can be bypassed by a sufficiently determined adversary.

Wrapping is not a panacea. The following sections describe some of the limitations of wrapping.

3.1. Limitations on Correct Behavior

Wrappers cannot stop all malfunctioning software. For example, if a WS were developed for a web server, the WS could reasonably prevent the web server from retrieving files that it did not have any need to access, but it could not prevent the web server from retrieving a configuration file and transmitting it to a remote site, which would be an undesirable action, but one not visible at the system call interface level. Similarly, if a wrapped program should allow certain users access to a particular file, but should deny access to other users, it is effectively impossible to enforce that constraint in the WS.

The premise of wrappers is that the WS should be much simpler than the software it wraps. If the WS needs to know as much and keep as much state as the application being wrapped, then the WS will be no simpler than the application. In this case, wrappers degenerate to a variation on N-version programming, which is both expensive and not particularly resilient against flaws.

3.2. Flaws in the Underlying Operating System

In the UNIX version of NAI Labs wrappers, the WEL and the WS run as part of the operating system. As a result, flaws in the wrapper can bring down the operating system. Additionally, flaws in the operating system can subvert the wrappers. For example, if there is a flaw that allows an application program to subvert or bypass the wrapper, the wrappers technology will be unable to stop it.

3.3. Limitations of Knowledge

Both the UNIX and NT versions of NAI Labs wrappers use a Wrapper Definition Language (WDL), which provides "characterization" of system calls (i.e., grouping related calls together). The characterization also names and types parameters. This characterization allows the WS developer to be ignorant of the system call details, and instead to focus on the classes of operations. However, if the characterization is incorrect (i.e., it misses one or more related system calls), then not all calls will be captured.

By contrast, the ISI wrappers require the developer to write a DLL that specifies each interface with its parameters. If the developer omits an interface (or makes an error in specifying parameters), the WS will not have the expected results. ISI wrappers do not provide a characterization capability, thus requiring the developer to enumerate each interface to be intercepted.

Both NAI Labs wrappers and ISI wrappers have capabilities for composition, although we did not use that facility in this effort.

In either case, though, the skill of the WS developer is critical in the quality of the resulting WS. For this reason, WSs for security critical systems like firewall proxies need to be vetted particularly carefully.

3.4. Total Wrapping

Because wrappers are just programs on a computer system, they are only as good as their protection. That is, even if a wrapper cannot be bypassed, if an attacker can modify or delete the WEL or WSs, then they can subvert the wrapper indirectly. For this reason, one of our efforts is to develop a "total wrapper" for Gauntlet. A total wrapper consists of a WS for the key parts of the system (e.g., for the proxies) plus a separate WS for everything else on the system that prevents interference with the WSs, the WEL, and the proxies themselves. The result is a system where an attacker who breaks through a proxy may run amok within the system, but will be unable to interfere with the wrappers or the proxies themselves.

For example, the total wrapper (also known as an interference preventer) might prevent anyone from binding the ports used by the wrapped proxies, modifying the WEL, the WSs, the proxy software, the configuration files used by the proxy, and shared libraries used by the proxy, etc. It must also constrain those programs with root privilege, limiting them to performing only those operations which will not interfere with the remaining wrappers.[§]

4. GAUNTLET WRAPPERS USING NAI LABS WRAPPERS

As described above, wrappers provide the means to constrain execution of arbitrary programs, including those running as root. In an application proxy-based firewall such as Gauntlet, proxies are started as root, perform key initialization functions (e.g., binding to low-numbered TCP ports), and then switch to an unprivileged user ID for "normal" operation. However, there are several risks remaining:

- If the proxy contains incorrect code that causes it to perform more than just binding its reserved port before relinquishing root permission, then it could perform arbitrary damage.
- If the proxy contains incorrect code that causes it to perform incorrect operations which are still allowed with its unprivileged ID, it might disclose information which should be restricted.
- If the proxy is vulnerable to buffer over-runs, it may be possible to cause it to perform undesired functions (e.g., creating a shell on the firewall machine).

These risks are exacerbated when proxies are written by programmers who are not familiar with the security

[§]This begs the question of how wrappers are installed in the first place. Initially, any program with root privilege can install a wrapper. When the wrapper is activated for a process, its first action is to disable the capability to install or otherwise modify the wrapper configuration. Thus, even if the process does not give up the root privilege, the wrapper will prevent itself from being subverted. This can be seen in the use of the *wrapper SetupAllowed* variable in Figure 3.

implications of proxy development. In particular, proxies written by end users (rather than firewall developers) may be more likely to have such flaws.

Proxies should be protected so they operate within a restricted subset of available system calls. The ideal subset should be the set of calls used by a proxy. However, without source code the only way to determine the system calls and arguments used is through a system call wrapper that captures the call profile, thus generating a report of what system calls are used by a particular program. While we had source code to Gauntlet available, we wanted to develop the wrapper as if source code were unavailable (in preparation for the experiment described in Section 5). The completeness of the system call set depends upon the ability of the person exercising the proxy to hit all code paths. In reality, this analysis will usually be incomplete because the conditions needed to exercise a code path will not be known.

Although the firewall's standard policy is to disallow that which is not explicitly allowed, this probably should not be the method used with system calls. Instead, it makes more sense to rely on the wrapper system call classification to, in general, allow or disallow entire classes of operations. Trying to create too precise a wrapper would only result in breaking the proxy, either because of previously unseen code paths or updated versions of the firewall. In cases where it is unclear from initial analysis whether a call or set of operations should be allowed, the wrapper could allow the operation but log the operation so that subsequent analysis can be done to define a more precise wrapper.

For example, well-behaved proxies may need to fork children (the Gauntlet HTTP proxy does this), but they should not need to exec other executables. In general, proxies should not need to write to the file system though specific proxies may need to. Preventing file system write operations where possible will prevent subverted proxies from damaging the file system. This includes both normal file writing and directory modifying operations such as link, unlink, and rmdir.

Most proxies need to read firewall-related configuration files. They may also need access to certain systemwide configuration files, which are usually in /etc. Even if they are not running in a chrooted environment, proxies can be restricted to a well-defined set of files they are allowed to read.

The very essence of a proxy is network I/O so they will need to perform socket-related system calls. Depending on the proxy, it may be possible to restrict the ports it is allowed to access so that a proxy cannot poke extra holes in the firewall.

Proxies do not need most “root” operations even if they are required to run as root. For those proxies that need root-only system calls, they should be restricted to only the needed calls and the arguments should be validated.

Our first step was to create wrappers for some of the more common firewall proxies, such as the HTTP, SMTP, and FTP proxies. In particular, the HTTP proxy is fairly complex, and as such is more likely than others to have security flaws. Figure 2 shows a wrapped Gauntlet with some existing and some new proxies wrapped.

The proxy wrappers are fairly simple. For example, the set of restrictions enforced for the HTTP proxy are:

- Disallow all system administration calls except those specifically exempted below.

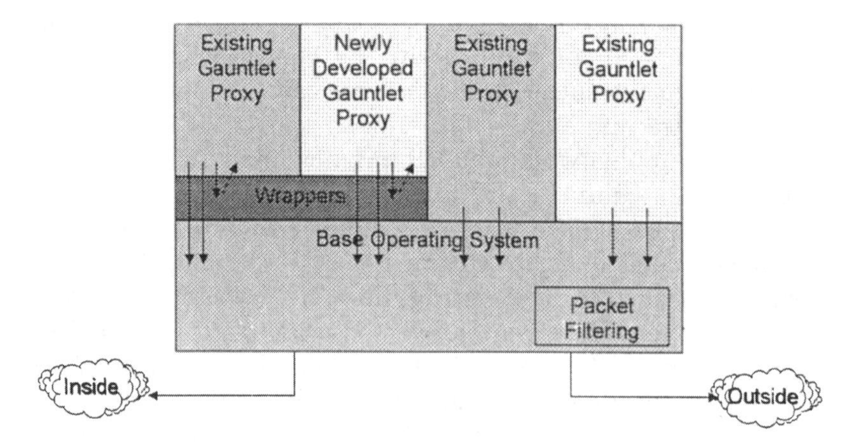

Figure 2 A wrapped Gauntlet.

- Allow setuid only to the value from the Gauntlet configuration (uucp by default).
- Allow setgrp only to the value from the Gauntlet configuration (6 by default).
- Disallow writing to the file system.
- Disallow file system operations like unlink, link, rmdir, etc.
- Allow reads only from the Gauntlet installation directory (/usr/local/etc).
- Allow bind only to socket configured by Gauntlet (80 by default).
- Fork is allowed but exec is not.

The resulting wrapper to implement this policy is also fairly simple, and is shown in Figure 3 (certain preliminary definitions have been omitted in the interest of space).

It is up to the author of the wrapper whether system calls should be allowed or denied by default; in this wrapper, the default is that system calls are allowed. Thus, there is a wrapper for the exec system call (to prevent its use), but not for the fork system call (which is by default allowed).

While there are good reasons to disallow system calls by default, protecting the “dangerous” calls and allowing all others is less likely to break existing code than the alternate approach.

The most interesting part of the wrapping exercise was testing. While we assume there are flaws in any proxy, we decided that it was easier (and perhaps more instructive), to deliberately introduce flaws, and then verify that the wrappers prevented a user from exploiting the flaws. Our main concern in deliberately introducing a flaw was the risk that it might somehow get checked into the main source tree, and become part of the shipping source code. To avoid this, we have taken steps to ensure that the modified source code is not provided to anyone in the product development organization.

Our deliberate flaws included introducing a capability for creating a root shell and binding to a disallowed port. Because the exec system call was blocked by the wrapper, the effort to

```
wrapper httpProxy {

    static int wrapperSetupAllowed;

    wr_activate () {
        // Prevents activation of other
        // wrappers once this one is started
        wrapperSetupAllowed = FALSE;
  }

    wr_duplicate () {
        // In case of fork(), prevents
        // child from replacing the wrapper
        wrapperSetupAllowed = FALSE;
    }

  WR_LOCAL::op{exec||exece}
    pre {
        wr_printf ("!!! Denying '%s'\n", $$);
        return WR_DENY | WR_BADPERM;
    };

  WR_LOCAL::op{bind}
    pre {
        struct sockaddr_in *sock = (struct sockaddr_in *) $addr;

        if (sock->sin_port != BIND_PORT) {
          wr_printf ("!!! Denying access to port %d\n", sock-
>sin_port);
          return WR_DENY | WR_BADPERM;
        }
    };

  WR_LOCAL::op{open || open64}
    pre {
        if (($flags & (O_RDWR | O_WRONLY)) == 0) {
          /* file is read only */
```

Figure 3 HTTP proxy wrapper.

```
        if (wr_strcmp ($path,
              "/etc/.name_service_door")
              != 0 &&
            wr_strcmp ($path,
              "/etc/.syslog_door") != 0 &&
            wr_strcmp ($path,
              "/etc/netconfig") != 0 &&
            wr_strcmp ($path,
              "/etc/hosts") != 0 &&
            wr_strcmp ($path,
              "/etc/nsswitch.conf") != 0 &&
            wr_strcmp ($path, "/dev/zero")
              != 0 &&
            wr_strcmp ($path, "/dev/conslog")
              != 0 &&
            wr_strncmp ($path,
              "/opt/SUNWspro/lib", 16)
              != 0 &&
            wr_strncmp ($path,
              "/usr/platform", 13) != 0 &&
            wr_strncmp ($path,
              "/usr/local/etc", 14) != 0 &&
            wr_strncmp ($path, "/usr/lib", 8)
              != 0 &&
            wr_strncmp ($path,
              "/usr/share/lib", 14) != 0)
          {
            wr_printf ("!!! Denying open for read of '%s'\n", $path);
            return WR_DENY | WR_BADPERM;
          }
        }
```

Figure 3 *(continued)*.

create the root shell failed. Because the bind system call was only allowed to the specified port, it failed as well.

As an adjunct to this effort, we developed a capability to generate alerts to an intrusion detection system (IDIP, 1998). The concept is that when the wrapper detects an attempt to

```
        else {
          /* open for write */
          if (wr_strcmp ($path,
                "/dev/conslog") != 0 &&
              wr_strcmp ($path, "/dev/null")
                != 0 &&
              wr_strncmp ($path,
                "/var/run/proxy.", 15) != 0)
          {
            wr_printf ("!!! Denying open for writing of '%s'\n",
$path);
             return WR_DENY | WR_BADPERM;
          }
        }
      }
    };

  WR_LOCAL::oattr{fileop}
    pre {
      if (wr_strcmp ($$, "open")   != 0 &&
          wr_strcmp ($$, "open64") != 0 &&
          wr_strcmp ($$, "stat")   != 0) {
        wr_printf ("!!! Denying fileop '%s'\n", $$);
        return WR_DENY | WR_BADPERM;
      }
    };

  WR_LOCAL::op{chroot}
    pre {
      if (wr_strcmp ($path, CHROOT_PATH) == 0) {
        wrapperSetupAllowed = TRUE;
```

Figure 3 (*continued*).

invoke a prohibited system call, this may indicate that a flaw has been found in the proxy and that the proxy is misbehaving. Of course, it may simply indicate that a code path is being exercised that was not found in developing the wrapper! Some system calls are more likely to indicate an attempted subversion than others. For example, for most proxies an exec call,

```
    }
    else {
      wr_printf ("!!! Denying chroot to '%s'\n", $path);
      return WR_DENY | WR_BADPERM;
    }
  };
WR_LOCAL::op{setuid}
  pre {
    if ($uid == SETUID_ID) {
      wrapperSetupAllowed = TRUE;
    }
    else {
      wr_printf ("!!! Denying setuid to %d\n", uid);
      return WR_DENY | WR_BADPERM;
    }
  };
WR_LOCAL::op{setgid}
  pre {
    if ($gid == SETGID_ID) {
      wrapperSetupAllowed = TRUE;
    }
    else {
      wr_printf ("!!! Denying setgid to %d\n", gid);
      return WR_DENY | WR_BADPERM;
    }
  };

// Catch all root operations and deny them
WR_LOCAL::oattr{rootop}
  pre {
    if (wrapperSetupAllowed)
      wrapperSetupAllowed = FALSE;
    else {
      return WR_DENY | WR_BADPERM;
    }
  };


}
```

Figure 3 *(continued).*

an attempt to bind ports other than that assigned to the proxy, or an attempt to read /etc/passwd is a sure sign of a break-in attempt, while attempts to read other files may be more likely to indicate an error in the wrapper. Interfacing with the intrusion detection system was done by writing records to the log file, and having a user space daemon read the log file and forward the relevant records to the intrusion detection system for processing. Because of timing concerns, it was not feasible to directly invoke the intrusion detection system from the wrapper running in the kernel.

The Gauntlet wrappers themselves are quite small. Not including the wrappers libraries, each wrapper is about 250 lines of commented code, thus lending themselves to easy inspection.

A related effort involved wrapping the Multi-Protocol Object Gateway (MPOG) (Lamperillo, submitted for publication), a gateway for CORBA, RMI, and DCOM traffic. Unlike traditional proxies, MPOG is written in Java, and runs under the Java Virtual Machine (JVM). Wrapping a JVM turned out to be difficult, because it performs many actions as part of startup that would ideally be blocked. The resulting wrapper, an excerpt of which is shown in Figure 4, is far more permissive than the corresponding HTTP proxy wrapper. We were quite surprised at some of the permissions necessary for proper operation of the JVM. For example, the JVM opens both /dev/null and /dev/zero for reading and writing. We would not have been surprised at opening /dev/null for writing, or /dev/zero for reading, but we were surprised that all combinations were required. We were disappointed that we had to open so many files for reading (e.g., /etc/hosts, /etc/resolv.conf) which may be of use to an attacker, should they be able to subvert the proxy. The number of files (and the specifics of the files) required by the JVM reinforced our belief that proxies should not be written in Java.

5. WRAPPING NONSOURCE FIREWALLS USING NAI LABS WRAPPERS

In Section 4 we described how we wrapped the Gauntlet firewall. Our next effort was to develop a wrapper for a fire-

```
wrapper mpog {
WR_LOCAL::op{exec||exece}
  pre {
    wr_log ("%s(%d): !!! Denying '%s'\n",
      _progname, _pid, $$);
    return WR_DENY | WR_BADPERM;
  };
WR_LOCAL::oattr{forkop}
  pre {
    wr_log ("%s(%d): !!! Denying '%s'\n",
      _progname, _pid, $$);
    return WR_DENY | WR_BADPERM;
  };
WR_LOCAL::op{open || open64}
  pre {
    if (($flags & (O_RDWR | O_WRONLY)) == 0) {
      /* file is read only */
      if (wr_strcmp ($path,
            "/etc/.name_service_door") != 0 &&
          wr_strcmp ($path,
            "/etc/.syslog_door") != 0 &&
          wr_strcmp ($path,
            "/etc/netconfig") != 0 &&
          wr_strcmp ($path,
            "/etc/hosts") != 0 &&
          wr_strcmp ($path,
            "/etc/mnttab") != 0 &&
          wr_strcmp ($path,
            "/etc/protocols") != 0 &&
          wr_strcmp ($path,
            "/etc/resolv.conf") != 0 &&
          wr_strcmp ($path,
            "/etc/nsswitch.conf") != 0 &&
          wr_strcmp ($path,
            "/dev/zero") != 0 &&
          wr_strcmp ($path,
```

Figure 4 MPOG proxy wrapper.

```
             "/dev/null") != 0 &&
           wr_strcmp ($path,
             "/dev/conslog") != 0 &&
           wr_strncmp ($path,
             "/opt/SUNWspro/lib", 16) != 0 &&
           wr_strncmp ($path,
             "/usr/platform", 13) != 0 &&
           wr_strncmp ($path,
             "/usr/local/etc", 14) != 0 &&
           wr_strncmp ($path,
             "/usr/java", 9) != 0 &&
           wr_strncmp ($path,
             "/usr/openwin", 12) != 0 &&
           wr_strncmp ($path,
             "/proc/", 6) != 0 &&
           wr_strncmp ($path, "./", 2) != 0 &&
           wr_strncmp ($path,
             "/usr/lib", 8) != 0 &&
           wr_strncmp ($path,
             "/export/home/guest/MPOG", 23) != 0 &&
           wr_strncmp ($path,
             "/usr/share/lib", 14) != 0)
       {
         wr_log ("%s(%d): !!! Denying open for read of '%s'\n",
                 _progname, _pid, $path);
         return WR_DENY | WR_BADPERM;
       }
     }
     else {
```

Figure 4 (*continued*).

wall for which we had no design information or source code. The goal was to see if the same techniques would work as they did for Gauntlet, and to determine whether we could add strength to competitive commercial products just as we did for Gauntlet.

```
      /* open for write */
      if (wr_strcmp ($path,
            "/dev/conslog") != 0 &&
          wr_strcmp ($path,
            "/dev/null") != 0 &&
           wr_strcmp ($path,
             "/dev/zero") != 0 &&
           wr_strcmp ($path,
             "/usr/local/etc/mp_orb_gw")
             != 0 &&
           wr_strncmp ($path,
             "/var/run/proxy.", 15) != 0)
      {
        wr_log ("%s(%d): !!! Denying open for writing of '%s'\n",
                _progname, _pid, $path);
        return WR_DENY | WR_BADPERM;
      }
    }
  };

WR_LOCAL::oattr{fileop}
  pre {
    if (
      wr_strcmp ($$, "open")    != 0 &&
      wr_strcmp ($$, "open64") != 0 &&
      wr_strcmp ($$, "resolvepath") != 0 &&
      wr_strcmp ($$, "pathconf") != 0 &&
      wr_strcmp ($$, "stat64") != 0 &&
      wr_strcmp ($$, "lstat64") != 0 &&
      wr_strcmp ($$, "stat")    != 0)
    {
```

Figure 4 *(continued).*

At the time this effort began, NAI Labs wrappers were available with FreeBSD 2.2.7 and Solaris 2.6. A version for Windows NT was under development, but was not stable enough for use when we began this experiment. A market survey revealed four firewall products that run on Solaris 2.6,

```
        wr_log ("%s(%d): !!! Denying fileop '%s'\n", _progname, _pid,
$$);
        return WR_DENY | WR_BADPERM;
      }
      else
        return WR_ALLOW;
      };

  WR_LOCAL::oattr{rootop}
    pre {
      wr_log ("%s(%d): !!! Denying call '%s'\n", _progname, _pid,
$$);
      return WR_DENY | WR_BADPERM;
    };

  }  // End of wrapper
```

Figure 4 (*continued*).

and none that run on FreeBSD. Of the four products, two vendors were unwilling to cooperate due to the competitive nature of Gauntlet with their products. We obtained both of the other products for our experiment.

Our results thus far have been discouraging. The first product we selected was SmallWorks NetGate. After installing and configuring NetGate, we realized that we had never specifically asked how the proxying is implemented. NetGate implements its proxies entirely within the kernel. Since our wrappers technology rely on intercepting system calls as they go from user state to system state, we were unable to develop any wrappers for it. Thus, there was no further effort on NetGate.

The second firewall we selected was Milkyway Networks SecurIT. As compared to Gauntlet, where each proxy implements its own access controls (and thus is wrapped separately), SecurIT uses a single program, guardian, to implement most of the security policies, including decision making based on IP addresses. Thus, a single wrapper could constrain many of the proxies. A second wrapper, which we called secureitproxy, serves to constrain many of the common proxies including ftp, telnet, and http. The operations allowed by the wrappers were roughly equivalent to those allowed for comparable Gauntlet proxies.

The result was that we were modestly successful in wrapping a firewall for which we did not have source code. We would like to (but have not yet) develop a method for measuring how much resilience was added by the wrappers. Also, because we have much less familiarity with the SecurIT product than Gauntlet, we would like to exercise the system more thoroughly, to see whether our wrapper improperly prohibited any activities which should have been allowed. The gross differences in firewall architectures were disappointing, in that they limited the usefulness of our approach.

6. WRAPPING GAUNTLET USING ISI WRAPPERS

Building on our results wrapping Gauntlet with NAI Labs wrappers, we decided to develop a set of wrappers for the Windows NT version of Gauntlet. For this effort, we used the ISI version of Wrappers (Balzer, 1999).

There are several key differences between NAI Labs' UNIX wrappers and ISI's NT wrappers. As noted in Section 3, ISI wrappers are implemented in user space, and as such can be bypassed by anyone who writes assembly or machine language code to make direct calls rather than going through the APIs. Thus, buffer over-run attacks will succeed against the ISI wrappers (since the over-run will typically operate by inserting machine code to perform its operations, which would not be stopped by the API-only view of the wrapper), while such attacks can be stopped by the NAI Labs' UNIX wrappers, which operate inside the kernel.[¶]

[¶]We note that NAI Labs' NT wrappers, which are not discussed in this paper, are subject to the same vulnerability as the ISI NT wrappers.

ISI wrappers are implemented as Dynamic Link Libraries (DLLs), and do not have a language for specifying wrappers akin to WDL. Windows NT has roughly 10 times more system call interfaces than a typical UNIX system. The lack of a characterization capability in ISI wrappers, together with the ten-fold increase in the number of interfaces available, made the wrapper definition much more time consuming. Additionally, the number and type of interfaces provided by NT changes radically from one version to another. Hence, there can be little confidence that NT wrappers are as thorough as the UNIX wrappers in constraining the behavior of misbehaving applications.

As in the case of the Gauntlet UNIX wrapping (Section 4) and SecurIT wrapping (Section 5), we focused on wrapping a small number of proxies (specifically, the ftp, http, and smtp proxies). While the policies we attempted to enforce were reasonably similar to the other firewalls, the implementations were rather different due to the differences between NT and UNIX. As an example, Figure 5 shows a comparison of the file portion of the wrapper.

Similarly, while Gauntlet UNIX stores most of its configuration data in files, Gauntlet NT stores its data in the Windows Registry. Thus, the wrapper needs to allow access to those portions of the Registry needed by the proxy.

An area of difficulty was process creation. While UNIX systems have fork and vfork system calls, NT systems have approximately 50 different APIs that can be used to create a new process. We determined experimentally that they seem to all call the same entry point in the kernel, although it is

Files	**Gauntlet UNIX**	**Gauntlet NT**
Hosts file	/etc/hosts	$WINDIR\System32\drivers\etc\hosts
Services file	/etc/services	$WINDIR\System32\drivers\etc\services
Netperm-table	/usr/local/etc/netperm-table	C:\Program Files\Network Associates\Gauntlet\netperm-table
Log file	/dev/log	\\.\pipe\Gauntlet\LogsService

(Where "$WINDIR" is the Windows NT directory, generally "C:\WINNT")

Figure 5 Comparison of Gauntlet UNIX and NT file restrictions.

difficult to know whether this is completely accurate. An attacker using a buffer over-run attack may cause execution of arbitrary machine instructions, which may invoke some of these alternate entry points (if they in fact exist). Since we can have no assurance that these alternate entry points are unused, it is safest to intercept them all. Also, Microsoft periodically changes the internal linkages between APIs and system calls, so there is no way to know whether this is a permanent situation. Thus, while we wrapped the single CreateProcessW API, we are unsure whether this would actually constrain a malicious proxy from creating other processes.

As noted above, each system call must be wrapped individually. Because of the lack of characterization, the wrapper developer must be intimately aware of the parameter types and orders, unlike with the NAI Labs wrapper technology. As a result, the development effort using ISI wrappers to wrap Gauntlet was much greater than using NAI Labs wrappers.

7. CURRENT STATUS AND FUTURE DIRECTIONS

Wrappers are not a universal solution, even for firewalls. Wrappers cannot solve certain types of security flaws, such as:

- An HTTP proxy that aims to block URLs based on the name of the URL (e.g., one containing the string "sex") or the content of the page.[1] Without replicating the logic of the proxy (which would mean duplicating the effort and complexity of the proxy in the wrapper), it is infeasible to verify the correct operation of the proxy against such requirements.

[1]Whether such "censorship" of URLs is good or bad, and the side effects of blind pattern matching, is beyond the scope of this paper.

- Proper implementation of an authentication method cannot be verified. For example, a wrapper could not protect against a malfunctioning telnet proxy that implements a one time password, but allows any users through even if the password does not match. Similarly, a wrapper cannot protect against a proxy that makes incorrect decisions based on SSL certificates.

In short, only those operations visible at the system call level with a minimum of understanding of the context can be protected using wrappers.

Our effort involved writing wrappers for firewalls that already exist, and in most cases do not have detailed designs. Thus, determining which system calls should be allowed and which should be rejected was a bottom-up effort, rather than top-down. We started by plugging in a logging wrapper which records each system call and its parameters, and then exercised the proxy. We undoubtedly did not exercise all call paths, and as a result it may be that the wrapper will refuse to allow certain (legal) operations that the proxy requires. This method of reverse engineering is timeconsuming and error prone, but it is the only effective method given the lack of design information.

The integration of NAI Labs wrappers with Gauntlet has been successfully demonstrated in DARPA's Technology Integration Center (TIC) to DARPA program managers as well as VIPs from other parts of government. The additional strength gained from the wrappers is a good example of "defense in depth",** one of the key approaches being used to provide security within the Department of Defense.

As part of the TIC demonstrations, government-sponsored "red teams" have attempted to subvert the wrappers. In fact, one of our motivations in developing the wrappers was a red team exercise in which the attackers

**The use of wrappers together with proxies is different from other uses of defense in depth, where multiple independent systems are used (e.g., a firewall with an intrusion detection system). However, it is an equally valid use of the concept: both the proxy and the wrapper are seeking to prevent an intruder from getting in to the internal system or into the firewall itself.

replaced the HTTP proxy with a booby-trapped copy which let them through at an appropriate time. Once we instituted the wrappers, the barrier was raised: the red team would have to not only replace the proxy, but also replace the wrapper with a corresponding wrapper that allowed the previously prohibited behavior to occur. We did not have the opportunity to run a second red team exercise after the wrappers were instituted, so we cannot determine how much additional resistance they provided.

The wrappers technology, including the firewall wrappers described in this paper can be downloaded from ftp.tislabs.com/pub/wrappers.

We have not performed any performance analysis of the wrapped firewalls. However, performance analysis of the base NAI Labs wrapper technology indicates that the load should be very small (i.e., probably in the range of 3–5%).

8. CONCLUSION

Wrappers can provide the ability to constrain the behavior of a firewall proxy, just as they can constrain the behavior of any other program in a general purpose computing system. Given the premise that wrappers are simpler (and hence more readily understandable) than the programs they wrap, a wrapper for a proxy adds a layer of protection even in cases where source code is available. In cases where source code is not available, wrappers may provide the only meaningful way to constrain the behavior of a proxy. By adding wrappers to a security device such as a firewall, we gain confidence that the firewall cannot be subverted, even if flaws in the proxies might otherwise allow such subversion to occur. Operating systems with more granular controls would render this technology unnecessary.

As one of the reviewers of this paper wrote, "... wrappers add security functionality to an operating system which helps improve firewalls... (which) implies that operating system controls are not very helpful—after all, not much about them has changed in 40 years."

ACKNOWLEDGEMENTS

This project would not have occurred without the encouragement of Sami Saydjari at DARPA. We also appreciate the encouragement from our colleagues on the DARPA Information Assurance program, including John Lowry, Andy Thompson, David Levin, and Gregg Schudel (all from Verizon, neé BBN) and Dan Schnackenberg (from Boeing). Finally, we appreciate the contributions of our many colleagues at NAI Labs, including Terry Benzel, Dale Johnson, Dan Sterne, Lee Badger, Mark Feldman, and Doug Kilpatrick.

We appreciate the comments from the anonymous reviewers, who helped improve this paper.

REFERENCES

Balzer, R., Goldman, N. (1999). Mediating connectors. In: Proceedings of the 19th IEEE International Conference on Distributed Computing Systems Workshop. Austin TX, May 31–June 5.

Fraser, T., Badger, L., Feldman, M. (1999). Hardening COTS software with generic software wrappers. In: Proceedings of the 1999 IEEE Symposium on Security and Privacy. Oakland CA, May.

IDIP (1998). Dynamic, Cooperating Boundary Controllers Final Technical Report. Boeing report D658–10822–1, Aug.

Jones, M. (1993). Interposition agents: transparently interposing user code at the system interface. ACM Symposium on Operating System Principles.

Lamperillo, G. Architecture and concepts of the MPOG. Unpublished manuscript.

Venema, W. (1992). TCP wrapper: network monitoring, access control, and booby traps. In: Proceedings of the 3rd UNIX Security Symposium. Baltimore, MD, Sept.

10

A Susceptible-Infected-Susceptible Model with Reintroduction for Computer Virus Epidemics

JOHN C. WIERMAN
Department of Applied Mathematics and Statistics, Johns Hopkins University, Baltimore, MD, USA

ABSTRACT

We consider a modification of the susceptible-infected-susceptible (SIS) epidemiological model for modeling the spread of computer viruses. The model includes a reintroduction or reinfection parameter, which models the re-release of a computer virus or the introduction of a new virus. We analyze the behavior of the model, and compare the results to previous research and to simulations.

1. INTRODUCTION

Cohen (1987) defined a computer virus as it is known today, as a program that can infect other programs by modifying them to include a possibly evolved copy of itself. It is distinguished by the properties of being self-replicating and of attaching itself to a host (program or other data object) for the purpose of concealing and transporting itself from one domain to another. In 1997, it was estimated that more than 10,000 viruses have appeared, and were being generated at a rate of roughly six per day (Kephart et al., 1997).

Computer viruses have cost billions of dollars, although the estimates are somewhat speculative. Total damage costs were reported to be $12.1 billion in 1999, $17.1 billion in 2000, and $10.7 billion for the first three quarters of 2001 (Abreu, 2001). In addition, users probably spend billions of dollars annually on anti-virus products and services. Consequently, methods to analyze, track, model, and protect against viruses are of considerable interest and importance.

Epidemic models for computer virus spread have been investigated since at least 1988. Murray (1988) appears to be the first to suggest the relationship between epidemiology and computer viruses. Although he did not propose any specific models, he pointed out analogies to some public health epidemiological defense strategies. Gleissner (1989) examined a model of computer virus spread on a multi-user system, but no allowance was made for the detection and removal of viruses or alerting other users to the presence of viruses.

More recently, a group at IBM Watson Research Center has investigated susceptible-infected-susceptible (SIS) models for computer virus spread. In Kephart and White (1991), they formulated a directed random graph model and studied its behavior via deterministic approximation, stochastic approximation, and simulation. In Kephart and White (1993) and Kephart et al. (1993), a combination of theory and observation led to a conclusion that computer viruses were much less prevalent than many have claimed, estimating that the number of infected machines is perhaps 3 or 4 per thousand PCs. They also claim that computer viruses are gradually becoming more prevalent, not because of any single viral strain, but because the number of viruses is growing with time.

This paper is an early report on a study of a modification of the SIS model begun in Wierman and Marchette (2004).

2. THE SIS MODEL

In the susceptible-infected-susceptible (SIS) model for a disease, each individual in the population is either infected (labeled I) or susceptible to infection (labeled S). When a susceptible individual becomes infected, it is immediately infectious. When an infected individual is cured, it is immediately susceptible again—no immunity is conferred. This is a homogeneous model, in which every infected individual has the opportunity to infect each susceptible, each infected has the same probability of being infected from each infected individual.

Ross (1915) introduced a deterministic version of the SIS model. A differential equation analysis led to a logistic curve for the infected proportion of the population, which predicts extinction of the infection when a basic reproductive ratio $R < 1$, and predicts a steady state endemic infection level if $R > 1$ whenever the initial infected proportion is positive.

A stochastic version of the SIS model, introduced by Weiss and Dishon (1971), is a continuous time Markov birth-and-death process used to model a variety of processes including epidemics (see Ball, 1999; Jacquez and Simon, 1993; Kryscio and Lefèvre, 1989; Nåsell, 1996; Nåsell, 1999), transmission of rumors (Bartholomew, 1976), and chemical reactions (Oppenheim et al., 1977). AQ1

The long-term behaviors of the deterministic and stochastic versions of the SIS model are quite different. While the deterministic model may settle into a steady state endemic infection level, in the stochastic SIS model of infection becomes extinct with probability 1, regardless of the parameters of the model. However, the time to extinction in the stochastic model can be extremely large, depending on the infection and cure rate parameters. The probability distribution of the number of infected individuals, during the long time until extinction, is sometimes approximated by the distribution conditioned upon non-extinction, which has been called the quasi-stationary distribution. The concept of the quasi-stationary distribution of a continuous-time Markov process was introduced by Darroch and Seneta (1967) for finite state-space Markov chains, and was first applied to epidemics by Kryscio and Lefèvre (1989), whose work was extended by Nåsell (1999) using asymptotic approximations.

3. COMPUTER VIRUS SIS MODELING

We consider the SIS model for a computer network of moderate size in an institution such as a corporation, university, or government agency. Since in the SIS model, when an individual is cured, it immediately becomes susceptible, the SIS model is particularly appropriate for computer virus models if one treats "virus infection" as infection by any virus. In this case, other than removing the computer from the network, any computer is, in principle, immediately susceptible, due to the introduction of new viruses. Due to the lack of spatial influences in human or animal populations, the homogeneity in the SIS model may make it more realistic for a network of computers than for biological populations.

For a network of n computers, due to homogeneity, we may formulate the SIS model as an $n+1$ state continuous time Markov process, where the states denote the number of computers infected by a virus. Let r denote the infection rate for any infected-susceptible pair, and let c denote the cure rate for any infected computer. The process is then a birth-and-death process with birth rates

$$\lambda_i = ri(n-i)$$

and death rates

$$\mu_i = ci,$$

for $i = 0, 1, 2, \ldots, n$.

Kephart and White studied a similar model both analytically and via simulation, finding that, under certain conditions, the infected population initially grew very rapidly. In fact, in the early stages of an epidemic, the process is well approximated by a branching process, sometimes called Kendall's approximation (Kendall, 1956), which has been rigorously established for many stochastic epidemic processes (See Ball, 1983a,b, Ball and Donnelly, 1995; Martin-Löf, 1986; Scalia-Tomba, 1985; von Bahr and Martin-Löf, 1980. After the rapid growth phase, the epidemic appeared to reach equilibrium. Since extinction is certain, this apparent equilibrium is temporary, where a quasi-stationary distribution persists for a long time until extinction finally occurs. The combination of the temporary equilibrium and eventual extinction simultaneously complicates the analysis of the model and makes it more interesting.

SIS WITH REINTRODUCTION

In this paper, we discuss a modification of the SIS model, in which we allow reintroduction of infection, at rate a, after extinction. The revised model is a birth-and-death process with rates

$$\lambda_0 = a, \quad \lambda_i = ri(n-i)$$

for $i = 1, 2, \ldots, n$, and

$$\mu_i = ci$$

for $i = 0, 1, 2, \ldots, n$.

The reintroduction feature may make the model more realistic for computer virus modeling. It corresponds to the possibility that the computer virus is "archived" either intentionally or unintentionally and reintroduced either unintentionally or maliciously at a later time. When considering infection to be more broadly defined than by a specific computer virus, but rather infection by any virus, then reintroduction can also correspond to the introduction of a new computer virus into the network.

The reintroduction feature is also mathematically convenient, because it eliminates the absorbing state, so the infected population size has a non-trivial limiting distribution. Since the temporary equilibrium in the classical SIS model persists for a long time, the limiting distribution in the SIS model with reintroduction may serve as an approximation to the quasi-stationary distribution that represents the temporary equilibrium. The two processes are identical until the classical SIS epidemic becomes extinct, after which the epidemic with reintroduction again becomes an active epidemic after a random waiting period, so the epidemic with reintroduction will always have a greater or equal number of infected computers than the classical SIS. On the other hand, the epidemic with reintroduction does occasionally become extinct for a short time, but then grows rapidly up to the temporary equilibrium again, by Kendall's approximation. Because of the relatively short time spent with low infected population sizes, the limiting distribution of the epidemic with reintroduction will attach only slightly less probability to the typical population sizes than the classical SIS temporary equilibrium distribution.

Thus, the limiting distribution can be expected to be a good approximation to the quasi-stationary distribution, but has the advantage that it can be described explicitly and studied analytically in terms of its parameters.

4. ANALYSIS AND RESULTS

The limiting distribution for the SIS model with reintroduction is given by the standard formula for the equilibrium distribution of a birth-and-death process (Ross, 1996, pp. 253–254):

$$P_0 = \left(1 + \sum_{i=1}^{n} \frac{\lambda_0 \lambda_1 \lambda_2 \cdots \lambda_{i-1}}{\mu_1 \mu_2 \mu_3 \cdots \mu_i}\right)^{-1}, \qquad P_k = P_0 \frac{\lambda_0 \lambda_1 \lambda_2 \cdots \lambda_{k-1}}{\mu_1 \mu_2 \mu_3 \cdots \mu_k}$$

Simplifying the factor in P_k produces

$$\frac{\lambda_0 \lambda_1 \lambda_2 \cdots \lambda_{k-1}}{\mu_1 \mu_2 \mu_3 \cdots \mu_k} = \frac{ar^{k-1}(n-1)!}{kc^k(n-k)!}$$

which with the help of Mathematica produces a closed form expression for the probabilities:

$$P_0 = \frac{c}{c + a_3F_1[\{1, 1, n - 1\}, \{2\}, -r/c]}$$

and

$$P_k = \frac{ar^{k-1}(n-1)!}{c^{k-1}k(n-k)!(c + a_3F_1[\{1, 1, n - 1\}, \{2\}, -r/c])}$$

for $k = 1,2,\ldots,n$, where ${}_3F_1$ is a generalized hypergeometric function (see Wolfram, 1996, pp. 750–751). While this formula gives an explicit exact solution for the limiting distribution, it is not easily interpreted and does not help us understand how the distribution changes as a function of the parameters.

In order to interpret the limiting distribution, we now discuss approximation of the limiting distribution for the number of infected computers for large network size n.

From the simplified expression for the factor in P_k, the limiting distribution depends on the cure and infection rates only through their ratio c/r. Since c is the cure rate per individual, while r is the infection rate for each infected-susceptible pair, it is appropriate to consider c/r to be a function of n to achieve a balance between the total infection and total cure rates for the network.

We now summarize the results of Wierman and Marchette (2004). Let X_n and $Y_n = n - X_n$ denote the random number of infected and uninfected computers, respectively, when the epidemic model is in the limiting distribution $\{P_{i,n}: i = 0, 1, 2,\ldots,n\}$, where the additional subscript n indicates dependence upon the population size.

If the cure rate is sufficiently small relative to the infection rate, one would expect it to be a rare event that any particular computer is not infected, suggesting an approximate Poisson distribution for the number of uninfected computers. In fact, this is the case:

If $\lim_{n\to\infty} \frac{c}{r} = b > 0$, then Y_n has a Poisson (b) distribution asymptotically.

This case corresponds to a particularly virulent infection, which is not consistent with observations of actual computer virus infections. However, we present it first because the analysis of this case provided the key to the analysis of other cases. Denoting the probability distribution of Y_n by $\{Q_{i,n}: i=0, 1, 2, \ldots, n\}$, we use the relation $Q_{i,n}=P_{n-i,n}$ to find the relationship

$$Q_{i,n} = \frac{n}{n-i}\left(\frac{c}{r}\right)^{i}\frac{1}{i!}Q_{0,n}$$

which, without the factor $n/(n-i)$, is satisfied by Poisson probabilities. Since for large n the factor $n/(n-i)$ approaches 1, the Poisson approximation is valid in the limit.

Since the standardized Poisson distribution is asymptotically normal as the mean tends to infinity, it is natural to expect that the number of uninfected computers has an approximate normal distribution in some cases when $\lambda_n = c/r \to \infty$ as $n \to \infty$. Wierman and Marchette (2002) show that

If $\lambda_n = o(n)$ or $\lambda_n = nd + o(n)$ where $d < 1$, then Y_n has an approximate Normal (λ_n, λ_n) distribution.

In these cases, of course, then:

X_n has an approximate normal $(n - \lambda_n, \lambda_n)$ distribution.

Although similar, somewhat different reasoning was needed to justify the result for the two different ranges for λ_n.

The range of normal approximation reaches only to the region where $\lambda_n = nd + o(n)$ where $d < 1$. Across this threshold value, when $d > 1$, the cure rate dominates, and the behavior is completely different:

If $\lambda_n = nd + o(n)$ where $d > 1$, then $\lim_{n\to\infty} P[X_n = k] = \frac{(a/k)(1/d)^{k-1}}{1-\log(1-1/d)}$ for $k = 0, 1, 2, \ldots$.

This is proved by directly taking limits of the frequency function and evaluating the normalizing factor using standard series convergence methods from calculus.

The limiting distribution is similar to, but differs slightly from the logarithmic distribution discussed in Johnson et al.

(1992), Chapter 7, given by

$$P[X = k] = \frac{\theta^k}{-k \log(1 - \theta)}$$

for $k = 1, 2, \ldots$, where $0 < \theta < 1$. The difference is due to the atom at zero and consequent different renormalization. The logarithmic distribution arises in a model strongly related to ours, being the steady state distribution of a birth and death process with rates $\lambda_i = \lambda i$ for all $i \geq 1$, and $\mu_i = \mu i$ for all $i > 1$ but $\mu_1 = 0$, a process appearing in Caraco (1979) in the context of animal group dynamics.

The authors are not aware of these logarithimic-type distributions arising in previous analyses of computer virus epidemic models. However, it appears to be particularly relevant for modeling the low observed levels of computer virus infection. As examples, the National Computer Security Association reported a computer virus infection rate of approximately 35 infections per 1000 computers per month in 1997, at www.webmastersecurity.com/ncsa97viruspreval-ence surveya.htm), and ICSA computer virus prevalence report, at www.truesecure.com/html/tspub/pdf/vps20001.pdf, reports slightly different rates.

Note that simulations in Wierman and Marchette (2004) show that the asymptotic approximations give a quite good fit for several choices of parameter values for population size $n = 100$.

5. DISCUSSION

We formulated and analyzed an elementary epidemic model for the prevalence of computer viruses in a homogeneous closed population of computers. Due to the homogeneity assumed in the model, it is more applicable to local networks or clusters of computers, where re-introduction may correspond to infection by a virus from outside the local network, rather than the entire Internet. With further development, epidemic model results may have value to a system administrator, for estimating the cost–benefit ratio of cleaning viruses and of purchasing anti-virus software.

For future research, a number of modifications might be incorporated to create models that are more realistic.

A model where the population increases with time might be more appropriate for the entire Internet. There might be different levels of subgroups of computers that communicate among themselves at different rates: For example, a three-level model could consist of departments within a company, companies, and the entire Internet. Different computers, or groups of computers (such as those with specific operating systems), might have different probabilities of becoming infected when exposed. Preventive action, such as the installation of anti-virus software, could be taken into account. In the epidemics literature, some of these effects have been taken into account, but not many combinations in any model for which rigorous mathematical analysis has been done.

ACKNOWLEDGMENTS

Prof. Wierman gratefully acknowledges research funding through the Navy-American Society for Engineering Education sabbatical program and the Acheson J. Duncan Fund for the Advancement of Research in Statistics, and the gracious hospitality of the Naval Surface Warfare Center at Dahlgren, VA, during his 2001–2002 sabbatical.

REFERENCES

Abreu, E. M. (2001). Computer virus costs reach $10.7B this year. Washington Post, Sep. 1, 2001, available at http://www.washtech.com/news/netarch/12267–1.html.

Ball, F. (1983a). The threshold behavior of epidemic models. *J. Appl. Prob.* 20:227–241.

Ball, F. (1983b). A threshold theorem for the Reed-Frost chain-binomial epidemic. *J. Appl. Prob.* 20:153–157.

Ball, F. (1999). Stochastic and deterministic models for SIS epidemics among a population partitioned into households. *Math Biosci.* 156:41–67.

Ball, F., Donnelly, P. (1995). Strong approximations for epidemic models. *Stoch. Proc. Appl.* 55:1–21.

Bartholomew, D. J. (1976). Continuous time diffusion models with random duration of interest. *J. Math. Sociol.* 4:187–199.

Caraco, T (1979). *Ecological Response of Animal Group Size Frequencies*. Fairland, MD: International Cooperative Publishing House, pp. 371–386.

Cohen, F. (1987). Computer viruses: theory and experiments. *Comput. Security* 6:22–35.

Darroch, J. N., Seneta, E. (1967). On quasi-stationary distributions in absorbing continuous-time finite markov chains. *J. Appl. Prob.* 4:192–196.

Gleissner, W. (1989). A mathematical theory for the spread of computer viruses. *Comput. Security* 8:35–41.

Jacquez, J. A., Simon, C. P. (1993). The stochastic SI model with recruitment and deaths. I: comparison with the closed SIS model. *Math. Biosci.* 117:77–125.

Johnson, N. L., Katz, S., Kemp, A. W (1992). *Univariate Discrete Distributions*. New York: John Wiley & Sons, Inc.

Kendall, D. G. (1956). Deterministic and stochastic epidemics in closed populations. In: Third Berkeley Symposium on Mathematical Statistics and Probability. vol. 4. University of California Press, pp. 149–165.

Kephart, J. O., White, S. R. (1991). Directed-graph epidemiological models of computer viruses. In: 1991 IEEE Computer Society Symposium on Research in Security and Privacy, pp. 343–359.

Kephart, J. O., White, S. R. (1993). Measuring and modelling computer virus prevalence. In: 1993 IEEE Computer Society Symposium on Research in Security and Privacy, pp. 2–15.

Kephart, J. O., White, S. R., Chess, D. M. (1993). Computers and epidemiology. *IEEE Spectrum* 30:20–26.

Kephart, J. O., Sorkin, G. B., Chess, D. M., White, S. R. (1997). *Fighting Computer Viruses*. Scientific American, Nov., pp. 88–93.

Kryscio, J., Lefévre, C. (1989). On the extinction of the SIS stochastic logistic epidemic. *J. Appl. Prob.* 27:685–694.

Martin-Löf, A. (1986). Symmetric sampling procedures, general epidemic processes, and their threshold limit theorems. *J. Appl. Prob.* 23:265–282.

Murray, W. (1988). The application of epidemiology to computer viruses. *Comput. Security* 7:139–150.

Nåsell, I. (1996). The quasi-stationary distribution of the closed endemic SIS model. *Adv. Appl. Prob.* 28:895–932.

Nåsell, I. (1999). On the time to extinction in recurrent epidemics. *J. Roy. Stat. Soc. B* 61:309–330.

Oppenheim, I., Shuler, K. E., Weiss, G. H. (1977). Stochastic theory of nonlinear rate processes with multiple stationary states. *Physica* 88:191–214.

Ross, R. (1915). Some a priori pathometric equations. *British Med. J.* 1:546.

Ross, S (1996). *Stochastic Processes*. John Wiley & Sons.

Scalia-Tomba, G. (1985). Asymptotic final size distribution for some chain-binomial processes. *Adv. Appl. Prob.* 17:477–495.

von Bahr, B., Martin-Löf, A. (1980). Threshold limit theorems for some epidemic processes. *Adv. Appl. Prob.* 12:319–349.

Weiss, G. H., Dishon, M. (1971). On the asymptotic behavior of the stochastic and deterministic models of an epidemic. *Math. Biosci.* 11:261–265.

Wierman, J. C., Marchette, D. J. (2004). Modeling computer virus prevalence with a susceptible-infected-susceptible model with reintroduction. *Comput Statist. Data Anal.* 45:3–23.

Wolfram, S. (1996). *The Mathematica Book*. 3rd ed. New York: Cambridge University Press.

11

Man vs. Machine—A Study of the Ability of Statistical Methodologies to Discern Human Generated ssh Traffic from Machine Generated scp Traffic

J. L. SOLKA and M. L. ADAMS
Naval Surface Warfare Center,
Dahlgren, VA, USA

E. J. WEGMAN
Center for Computational Statistics,
George Mason University,
Fairfax, VA, USA

ABSTRACT

This paper discusses our recent results in the classification of human-based ssh traffic as compared to machine-based scp traffic. Since both of these services, ssh and scp, use the same port, port 22, this classification problem occurs within a quite natural framework. Results that illustrate an exploratory analysis of the data will be presented along with some preliminary classification results.

1. INTRODUCTION

The attacks of September 11th have demonstrated to the United States of America and to the world the vulnerability of a country's infrastructure to attack. A country's infrastructure includes the traditional items such as transportation, communication, and energy facilities. These types of traditional infrastructure play an important role in a country, however, a more important role may be played by the computer network infrastructure of the country.

These cyber infrastructure attacks are perpetuated by a variety of individuals. These individuals run the gambit from bored teenagers to people who are under the control of organization which are only loosely coupled to state such as the Al Quaeda and can even include attacks that are sponsored by specific hostile countries.

Some of these attacks require human intervention, while many of the attacks specifically utilize automated tools in order to ascertain the vulnerabilities and attack particular systems. We as defenders of these systems would like to be able to tell whether the activity on a system has been under the control of a human being or machine. This paper attempts to take a small step toward this capability by exploring the capability to discern between secure shell (ssh) traffic and secure copy (scp) initiated traffic. In the first case, the data stream produced during the interaction is totally under the control of the human operator. In the second case, the data transference is initially human initiated but is then ultimately under the control of the computer.

The first section of this paper provides a little additional discussion as to the problem at hand. The next section examines some of the statistical problems of interest that are resident within the ssh vs. scp problem. The next section provides some of the results that we have obtained during our initial analysis of the problem and the final section besides summarizing provides a brief discussion of our future plans for continued research in this area.

2. BACKGROUND

2.1. Cyber Infrastructure as a Growing Concern

The computer network infrastructure has become ingrained in many if not all of the other infrastructure components. Many of the power plant systems, for example, are under the control of supervisory control and data acquisition (SCADA) systems. These SCADA systems provide a clear-cut avenue for cyberspace compromises to propagate into our everyday world with potentially dire consequences.

The banking sectors have also become dependent on the successful functioning of our cyber infrastructure. Cleverly planned attacks could disrupt the delicate flows of capability within the banking sector. Finally, we note that the commercial sector in general has become dependent on the unfettered flow of information. Previous distributed denial of service attacks have demonstrated the extreme vulnerability of the on-line commercial sector to attacks which seek to deny customers access to their websites.

2.2. Statistical Problems of Interest in the Cyber Infrastructure Arena

There are numerous problems of interest to the statistical community that are resident within the infrastructure protection area. Some of the relevant statistical techniques include exploratory data analysis, clustering, discriminant analysis, and visualization. The focus of our analysis within is in the area of exploratory data analysis and visualization with an eye toward discriminant analysis.

2.3. The ssh vs. scp Problem

Constraints on the length of this paper prevent us from providing a complete discussion of the inner workings of the ssh or scp service. The reader is referred to Stevens (1994) for a wonderful treatment as to the intricacies of Internet packet-based communications. scp and ssh are two services that can be configured to run on a platform in a server or client mode. ssh is essentially a telnet-type program where the data stream is subject to an encryption process. Similarly, scp is a file transfer or ftp type of program where the file transfer stream is also subjected to an encryption process. Typically, one machine is set up as a ssh/scp server and then a group of other machines, clients, run a client program that allows them to access the ssh or scp capability of the server. In this manner, the client can login to the server and run a typical terminal session or even tunnel various application through the ssh "pipe".

The part of this process that is relevant to our discussions is the fact that we are attempting our analysis not on each of the single packets that are exchanged during the session. Each session or interaction between two computers is based on the exchange of a sequence of packets. We have chosen from the onset of our analysis to work with sessions rather than individual packets. It would be virtually impossible to discern the difference between a single ssh and scp packet.

The ssh and scp services are typically configured so that they both run on port 22 of the server machine. In this manner packets associated with either scp or ssh sessions on the server machine would both be transmitted on an identical port. This fact is the focus of our analysis in that we are interested in rather one can distinguish between ssh and scp traffic on this port.

3. RESULTS

Here we describe our experimental protocol for data collection/processing, provide our results, and discuss them.

3.1. Data Collection Process

We initially planned to collect data using a server running scp and ssh on port 22 and then distinguish among the two services using user provided log files. It is also important to note that our facility really supports a single user per machine and hence we also have the capability to distinguish a particular user based on their IP address. This approach proved untenable in that each of the particular machines on our network presents a slightly different clock time due to different rates of central processing unit (cpu) clock drift on the system.

We addressed this data collection problem by modifying the configuration of the server in order to run ssh on port 76 and scp on port 77. Each user participating in the study then aliased their scp and ssh commands so that they accessed these nonstandard ports on our server. In this way we could trivially disambiguate the scp and ssh traffic.

3.2. Session Tracking Process

As mentioned in the previous section we do not use the raw packet signatures in our scp/ssh recognition scheme but rather use information/statistics computed on sessions. A session for our purposes can loosely be defined as a set of packet exchanges between a user on a particular client machine and our preconfigured server from initiation of the connection to the final termination of the session.

We used an in-house developed package designated as TRACKER or XTRACKER to take a sequence of packets associated with a particular session and assemble them into the associated session so that we can compute various statistics on the session.

3.3. Session Statistics Calculation

The features extracted for each session are as follows: day of the week, year, month, date, seconds since the beginning of time, source IP address, source port, destination IP address, destination port, number of packets transmitted during the session, number of packets into the server, number of packets out from the server, number of data packets (nonheader packets) transmitted, number of data bytes transmitted, duration of the session, and status of the session. A few minor clarifications about these features are in order. First, the beginning of time is measured from January 1, 1970, second the source port is the port on the client machine that the session was initiated from (usually a random high numbered port), the destination port is of course 76 or 77 depending on the service being requested, header packets are those portion of the datagram that contain some of the nondata type information associated with the packet, and status of the session is a flag that indicates the final status of the session. The status of the session flag is currently not employed during our analysis. The remainder of the features were chosen based on the current capabilities of the TRACKER/XTRACKER programs and our intuition regarding which features would be more useful in our quest to distinguish between scp and ssh traffic.

3.4. Visual Discernment of scp/ssh Class Structure

In this section we examine some of the various features that we have examined in the scp vs. ssh discriminant analysis problem. The first two-dimensional feature pair are *data packets* and *total packets*. The reasoning behind this feature pair is that the ratio of data packets to total packets for scp sessions should be greater than the ratio of data packets to total packets for the ssh sessions. In Figure 1, we plot number of session data packets vs. number of total packets for the ssh and scp sessions. The ssh sessions are labeled by red plus signs while the scp sessions are labeled by blue circles. A visual assessment of this plot would suggest that there would be little discriminant utility in this particular feature pair. Visual assessment of feature utility is a very tenuous process and we have chosen to also evaluate each feature pair using a simple classification scheme. We have chosen to use a one-nearest neighbor classifier and quantified the classifier performance using a standard leave one out cross-validation scheme. The reader is referred to Duda et al. (2000) for a discussion of the single nearest neighbor classifier and the cross-validation procedure. For this particular feature pair, the estimated performance figure of merit is a probability of correct classification of around 0.76. As will be revealed shortly,

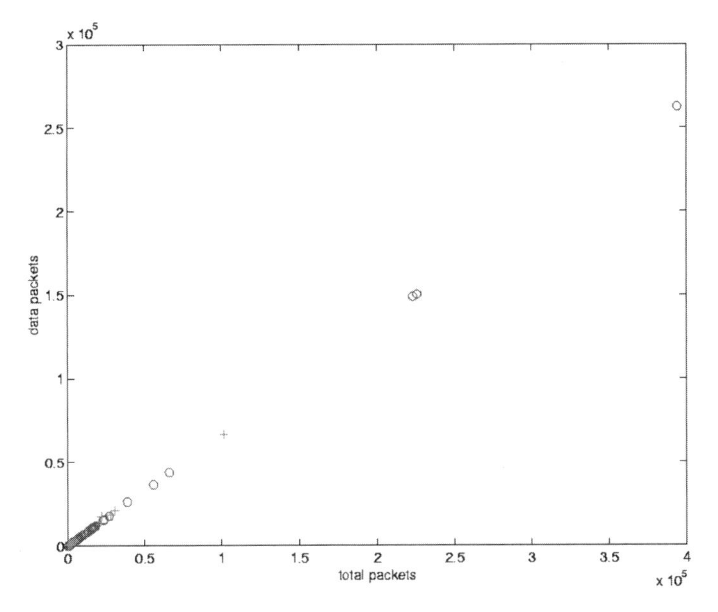

Figure 1 Number of data packets vs. number of total packets where a blue o indicates an scp session and red + indicates a ssh session.

this level of performance pales in comparison with that obtained with some of the other feature pairs.

The next feature pair is the absolute value of the difference between the number of data packets out from the server and the number of data packets into the server along with the total number of data packets. The justification of the use of these features is as follows. Consider an scp session wherein a user copies a large file from the server to his client machine. The actual initiation of the scp procedure involves many fewer packets into the server than the number of packets that are transferred from the server. In this sense one would expect the ratio of the absolute value of the difference in the number of data packets out—the number of data packets into the server to the total number of data packets would be larger in the case of the scp sessions. This expectation is supported by Figure 2. This feature pair seems to offer more hope with regards to distinguishing between the two classes and this

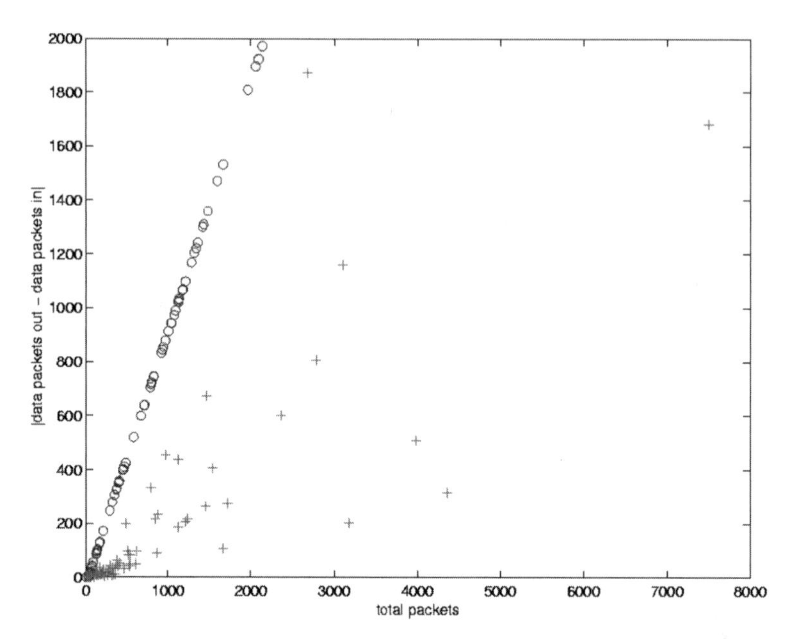

Figure 2 Absolute value of the number of data packets out of the server minus the number of data packets into the server vs. the total number of data packets. The scp observations are plotted as blue circles while the ssh observations are plotted as red plus signs.

hypothesis is supported by the measured single nearest neighbor cross-validated performance figure which is 0.95.

The next figure combination that we consider is the natural logarithm of the number of data packets out of from the server in conjunction with the log of the number of data packets into the server. We expect this feature to provide a high discriminatory power because in an ssh session there is more of a one to one exchange of packets between the client and server machines while in a scp session there is a more dichotomous relationship. An initial set of client provided packets leads to a large number of packets solicited from the server. For example, if one were using scp to copy a large file from the server then there would be many more packets out of the server than into the server. This speaks to the fact that this feature is directional in nature but this does not in practice present a problem to the analyst in that the direction of the packet flow is known as part of the session information.

Figure 3 presents a plot of the natural log of the number of data packets out from the server vs. the natural log of the number of data packets into the server. The scp sessions are plotted as blue circles in the plot while the ssh sessions appear as red plus signs. One can distinguish a clear separation between the observations associated with these two types of sessions and this perceived separation is supported by the estimated cross-validated probability of single nearest neighbor classification performance of 0.98.

We do note that the discriminatory power decreases in the case where there are not many packets associated with

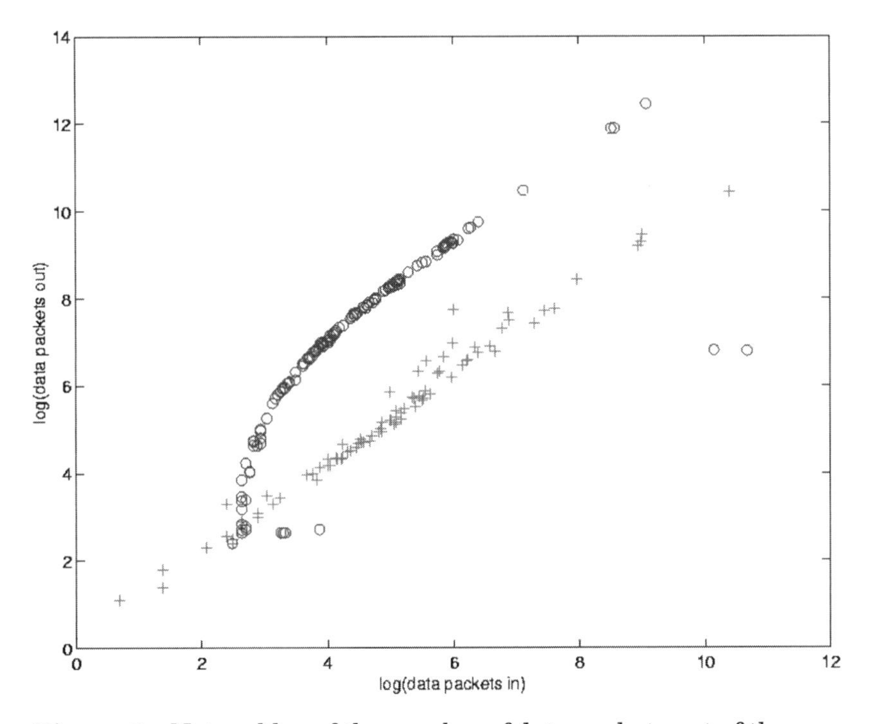

Figure 3 Natural log of the number of data packets out of the server vs. the natural log of the number of data packets into the server. The scp sessions are plotted as blue circles while the ssh sessions are plotted as red plus signs.

the session. An scp example of this would be when a user copies a very small file. In this case, the plot for the scp observations and the ssh observations seems to convergence at a v-shaped structure. The close proximity of the scp and ssh observations in this case clearly indicates the lack of discriminatory power and this behavior is in keeping with our intuition.

The next feature pair that we will examine is the natural log of the number of data bytes along with the natural log of the duration of the session. We would expect given the same duration for the session that the number of data bytes transferred during a scp session would be much higher than the number of data bytes transferred during a ssh session. A human typist would not be expected to be able to keep up with the associated transfer rates that one would expect to exist

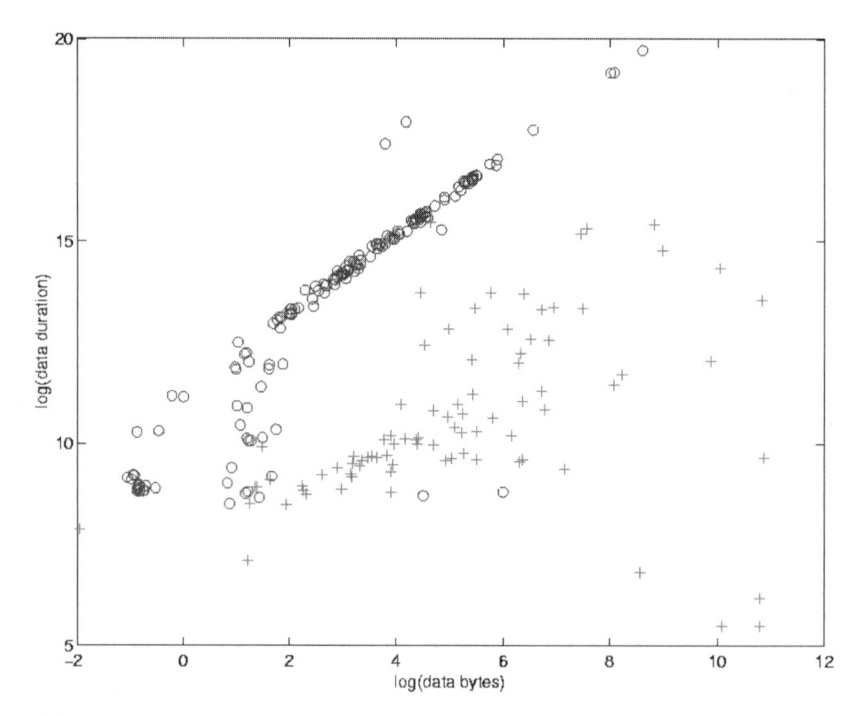

Figure 4 Natural log of the number of data bytes vs. the natural log of the duration of the session. The scp sessions are plotted as blue circles while the ssh sessions are plotted as red plus signs.

during an scp session. This assumes of course that the user is not involved in tunneling any sort of traffic through the ssh pipe. In this case one would expect this measured feature pair to fall somewhere between the pure ssh xterm-type session and the scp totally machine controlled session. Figure 4 presents a plot of the natural logarithm of the number of data bytes vs. the natural logarithm of the duration of the session. Once again we have plotted the scp observations as blue circles and have plotted the ssh observations as red plus signs. As expected, the scp observations in general lie above the ssh observations as expected. This observation is supported by the measured performance metric of 0.95.

We point out for completeness that this feature suffers from the same lack of discriminatory power in the case of short duration sessions. It is not surprising in this case that the scp observations and the ssh observations are very close. In this case the number of packets transferred as part of an scp session and the number of packets transferred as part of an scp session are very close.

4. CONCLUSIONS

We have presented a new approach to the discernment of scp and ssh sessions. We have chosen to utilize a number of session-based features for this preliminary exploratory data and discriminant analysis study. We have examined a number of two-dimensional orthogonal projections of our extracted feature set. We have provided simple scatter plots and one nearest neighbor cross-validated classifier performance for each of the two dimensional feature combinations. These preliminary results suggest that the natural logarithm of the number of data packets out along with the natural logarithm of the number of data packets in is the optimal feature pair among those two-dimensional pairs studied. This provided a measured figure of merit of 0.98.

There are numerous things that were not done as part of this study. First, it would make sense to quantify performance using the full set of extracted features or at least those features that are appropriate to include. Second, it would make sense to examine linear combinations of the features. The astute reader might wonder why we have not provided results related to the Fisher linear discriminant procedure, for example. It would also make sense to examine the features within a hyperdimensional framework such as parallel coordinates. In this manner we could ascertain the overlap between the observations in the full feature space and possibly rapidly determine those feature combinations that would possess the most discriminatory power. This parallel coordinates framework could be integrated with a procedure to explore linear projections such as the grand tour of Asimov. This would provide a convenient framework to identify fortuitous linear combinations of the features.

There are two other issues that we have not discussed due to the page limitations of this paper. The first is the problem of distinguishing between various users based on their ssh or scp utilization patterns. The second issue is the detection of tunneled applications. One can ssh into a server and then tunnel any X-Windows type of application. Although this is not normally done, one could tunnel a web session or text editor session such as emacs. The detection and classification of these types of tunneled applications are also of interest but will have to be relegated to future investigations.

We finally note that this analysis has been based on a very small data set. The data set suffers from not only lack of cardinality but also from bias in that the distribution of observations across IP address (user) is not uniform. We hope to address these experimental design issues with future data collections.

ACKNOWLEDGMENTS

The first two authors (JLS and MLA) would like to acknowledge the support of the NSWCDD ILIR Program along with the support of the Missile Defense Agency. The work of the third author (EJW) was completed under the sponsorship of the Air Force Office of Scientific Research under the contract F49620-01-1-0274 and the Defense Advanced Research Projects Agency through cooperative agreement 8105-48267 with Johns Hopkins University. The authors would also like to thank Dr. David Marchette for initially suggesting this effort, and for his many insightful comments during the experimental design and analysis phase of this research. Finally, the authors would like to thank Mr. Don Talsma for providing us access to the XTRACKER AND TRACKER software.

REFERENCES

Duda, R. O., Hart, P. E., and Stork, D. G. (2000). *Pattern Classification*. New York: Wiley-Interscience.

Stevens, W. R. (1994). *The Protocols*. TCP/IP Illustrated, Vol. 1. Reading, MA: Addison-Wesley.

12

Passive Detection of Denial of Service Attacks on the Internet

DAVID J. MARCHETTE
Naval Surface Warfare Center,
Dahlgren, VA, USA

1. INTRODUCTION

The public Internet is a critical component of the information infrastructure supporting finance, commerce, and civil and national defense. Denial of service attacks on major Internet sites, both the direct effect on the attacked sites and the indirect collateral effects on the Internet as a whole do considerable financial damage on a regular basis. Denial of service attacks could be a part of a concerted attack on the flow of information. This coupled with a physical attack of some kind poses a substantial threat.

Suppose, for example, that there is a coordinated denial of service attack (using one of a selection of freely available tools) on the public banking access sites of the 10 largest US banks (or the largest stock trading sites). Financial institutions are reluctant to share information, so it might take a while (hours or days) to sort out the size and the scope of the attack, or to even find out that the attack took place. A method of determining the scope of the attacks without relying on self-reporting is clearly needed.

In order to understand the typical denial of service attack, it is necessary to understand the workings of the TCP protocol. The protocol has two important extensions beyond the basic IP protocol. First, it (like UDP) implements the concept of "ports", which can be thought of as a local extension of the IP address by two bytes. Each session to or from an application is assigned a unique port, and the source and destination port pairing is used to disambiguate multiple ongoing sessions between machines. Second, TCP implements a two-way connection scheme, via the use of flags in the header. Thus, a typical TCP session is initiated as follows.

The client sends a packet with the SYN (synchronize) flag set, indicating that it wants to communicate. The destination port is set to the port of the application (for instance, port 80 is used primarily for web traffic); the source port is set arbitrarily, uniquely to this session, and is used to identify this specific session. The server responds (if it wishes to accept the connection) with a SYN/ACK packet (both the synchronize and acknowledgment flags set), and finally the client responds with an ACK packet (only the ACK flag set). This three-way handshake sets up the connection, and allows two-way communication between the hosts to proceed. This discussion has been simplified. There are sequence numbers used to ensure the ordering of the packets, and there are flags that control and end the communications after the initial connection is set up, but this is sufficient to understand what follows. Details can be found in Stevens (1994) or Marchette (2001).

The basic idea of most denial of service attacks is to flood a computer with bogus requests, or otherwise cause it to devote resources to the attack at the expense of the legitimate users of the system. A classic in this genre is the SYN flood.

The attacker sends SYN packets requesting a connection, but never completes the handshake. One way to do this is to set the source IP address to a nonexistent address (this process of changing the source address is called "spoofing" the address). For each SYN packet, the victim computer allocates a session and waits a certain amount of time before "timing out" and releasing the session. If enough of these "bogus" SYN packets are sent, all the available sessions are devoted to processing the attack, and no legitimate users can connect to the machine.

A related attack is to send packets that are out of sequence, or errors, forcing the victim computer to spend time handling the errors. For example, if a SYN/ACK packet is sent without having received an initiating SYN packet, the destination computer generates and sends an RST (reset) packet. If the attacker can arrange to have millions of SYN/ACK packets sent, the victim computer will spend all its resources handling these errors, thus denying service to legitimate users. One way to arrange this, is through a distributed denial of service tool, such as trinoo or TFN2k. These tools compromise a set of computers, dispersed across the IP address space, then use these "intermediate victims" to send thousands of packets to the intended victim. Each packet is crafted to have a random (spoofed) source IP address, so the attacking machines cannot be identified. See Marchette (2001), Chen (2001), and Northcutt et al. (2001) for descriptions of some distributed denial of service attacks.

The result of such an attack is a number of reset (or other) packets appearing at random sites around the Internet, with no obvious session or initiating packets to explain them. See Figure 1. This is used by Moore et al. (2001) to estimate the number of denial of service attacks during three one week periods, by counting how many unsolicited packets are seen addressed to one of the 2^{24} possible IP addresses they monitored.

Unsolicited packets resulting from a denial of service attack are called "backscatter" packets. It is these packets that we will use to investigate the denial of service attacks on the Internet. Note that this will be a completely passive

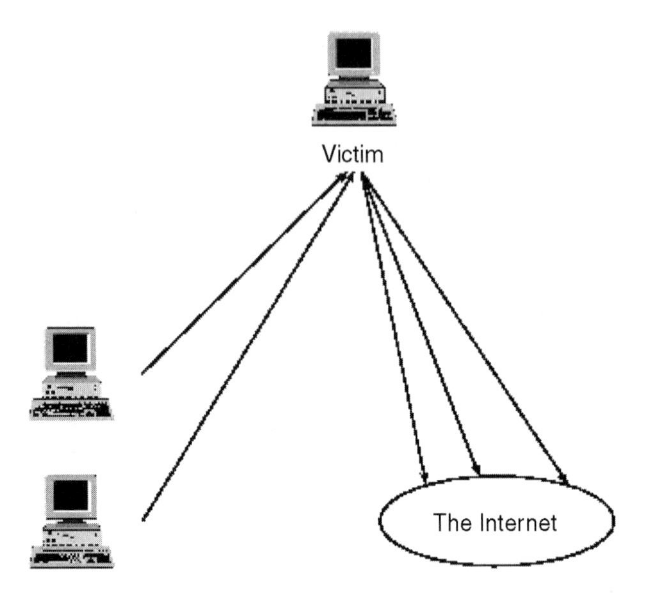

Figure 1 Backscatter from a denial of service attack. Packets are sent to the victim from one or more attackers. These packets have spoofed IP addresses, which cause the victim's response to be sent to random addresses within the Internet.

analysis: we will simply collect packets that are sent to our network. There is no requirement for cooperation from the victim(s), nor even any necessity that anyone outside our network know about the monitoring. Finally, our monitoring will not put any additional load on the network, and for those sites that already monitor the traffic in and out of the network, no new data need be collected.

2. ANALYSIS

Following Moore et al. (2001), we can compute some of the probabilities of detection needed to analyze backscatter packets. Assume the spoofed IP addresses are generated randomly, uniformly on all 2^{32} addresses, and independently.

Assume there are m packets sent in an attack on a given victim. If we monitor all packets to IP addresses, then it is easy to see that the probability of detecting an attack is

$$P[\text{detect attack}] = 1 - \left(1 - \frac{n}{2^{32}}\right)^m \tag{1}$$

From this, one obtains the result that the expected number of backscatter packets we detect is

$$\frac{nm}{2^{32}} \tag{2}$$

We would like to determine how many packets were originally sent. This will give an estimate of the severity of the attack, and might allow us to infer whether the attack was likely to have been mounted by multiple attackers, for example, through a distributed denial of service tool. To do this, note that the probability of seeing exactly j packets, under our independence assumption, is

$$P[j \text{ packets}] = \binom{m}{j}\left(\frac{n}{2^{32}}\right)^j\left(1 - \frac{n}{2^{32}}\right)^{m-j} \tag{3}$$

The maximum likelihood estimate for m, using Eq. (2.3), is

$$\hat{m} = \left\lfloor \frac{j2^{32}}{n} \right\rfloor \tag{4}$$

Thus, if we see j packets, we can use Eq. (4) to estimate the size of the attack.

Note that if the attacker chooses from a subset of all possible IP addresses, say of size N, then we must replace 2^{32} in Eqs. (1)–(4) with N. These equations, then, under the assumptions of uniformity and independence, allow us to estimate the original size of the attack from the packets we see at our network, assuming we know the original size of the pool of IP addresses from which they are selected.

There is also the question of determining the number of attacks. Moore et al. (2001) do this by defining an attack as a series of packets with a maximum interpacket gap less than a fixed value. The idea being that if there is a long enough gap between packets, then it is reasonable to assume that these correspond to different attacks. They then count the number of attacks they detect, and report over 12,000 attacks in the three week period they investigated. (In comparing this number with the ones we report below, it is important to realize that Moore et al. were considering all denial of service attacks which produce backscatter, while we will be considering only that subset which produces SYN/ACK packets, and of these, only attacks against web servers.)

All of this assumes that all attacks generate packets to the network monitored. We will assume that all attack packets generate backscatter (until the machine ceases to function), ignoring issues such as filtering firewalls or other kinds of mechanisms that may block either the attack or the backscatter packets. If our monitored network (n IP addresses) is sufficiently small, and a sufficiently small number of packets are sent in the attack, there is a reasonable probability that we will receive no packets.

A note on the categorical nature of the data is in order. Technically, IP addresses are categorical, and certainly they are discrete. However, under the null hypothesis, they are generated as 32-bit integers, and so any analysis that treats them as such is valid. The categorical nature of the data can be ignored for the purposes of these analyses.

Several calculations are possible to determine whether the assumptions are valid. For example, Moore et al. (2001) suggest using the Anderson–Darling test of uniformity to test that the IP addresses are in fact uniformly distributed. We will discuss this further below. This of course assumes we know how many IP addresses are in the complete pool. A perusal of the attack code available on the Internet shows that the tools often allow the user to choose which octets of the IP address to randomize, thus reducing the pool. Assume the pool contains N addresses, and we monitor n addresses, and that the attack packets are sent every t seconds. Then if we knew how long a gap one should expect to see between detected packets, one could use this to estimate N, and thus be able to use the equations above for the estimate of the size of the attack. This would also be important for the determination of a definition of attack, as one would want the gap to be many standard deviations larger than this expected delay. The calculation is straightforward. The expected number of attack packets between two detected packets (assuming independence) is

$$\sum_{s=1}^{n}\left(1-\frac{n}{N}\right)^{s-1}\frac{n}{N}s \approx \frac{N}{n}$$

The variance of the number of packets between two detected packets is

$$\sum_{s=1}^{n}\left(1-\frac{n}{N}\right)^{s-1}\frac{n}{N}s^2 - \left(\sum_{s=1}^{n}\left(1-\frac{n}{N}\right)^{s-1}\frac{n}{N}s\right)^2 \approx \frac{N(N-n)}{n^2}$$

So, from this we see that we expect a gap of around tN/n seconds between packets from a given victim. For example, if an attacker sends 100 packets per second, and one monitors 2^{24} addresses, one expects to see a new packet about every 2.5 sec, and a spread of three standard deviations gives a 10 sec gap. This rate of attack is quite low (Moore et al. 2001 claim intensities of as large as 600,000 packets per second), but this is only for illustration's sake (even at this rate, a SYN flood can be quite effective). Similar calculations can easily be done for other values of n, N, and t.

All these calculations have been predicated on the attacker choosing randomly from 2^{32} possible IP addresses. Many attack tools choose from a subset of these, such as only selecting octets from the range 1 to 254 (which avoids "broadcast" packets which are routed to all machines on the subnet). This can be easily incorporated in the above analysis, by replacing the 2^{32} by the appropriate number.

3. EXPERIMENTAL RESULTS

To determine the extent that the assumptions of the theory are met, we consider a data set taken from a network of 2^{16} IP addresses. The data consist of unsolicited SYN/ACK packets received during two periods: April 4, 2001 – Jul 16, 2001 and September 1, 2001 – Jan 31, 2002. During these periods there were times when the sensor was down, for a total of 210 hr. The full data set consisted of 5842 hr. We refer to the network on which the data were collected as the "protected network" throughout this discussion.

Missing data bring up one of the practical issues in a study of this kind. The protected network is a working network with a moderate load, and so there is the problem of determining which packets were solicited and which were not. This is exacerbated if there are packets that were not captured by the sensor, either because it was unable to handle the load or because the sensor was down. With SYN/ACK packets, we need to know if the SYN packet was sent. If it was, and the sensor failed to capture it, we will notice further packets (ACKs, PUSHs, etc.), and can therefore determine that the SYN/ACK is a part of a legitimate session, and therefore not backscatter.

The other packet type that is commonly generated in denial of service is a RST (reset flag set) packet. These are quite common in normal operation, and can occur quite a long time after the (legitimate) session has ceased to generate packets. Note that Moore et al. (2001) had the luxury of a large unpopulated network for their data collection. We do not have this luxury, and so the problem of separating backscatter packets from legitimate packets (or those otherwise unrelated to denial of service attacks) is one that must be addressed. For this reason, we focus on SYN/ACK packets in this chapter.

3.1. THE DATA

In order to avoid the gaps in our data collection, we broke the data into eight subsets, as depicted in Table 1. These are named according to the last month in which data was collected for that subset. As will be seen, this split was not perfect, as there were still a few gaps within one of the larger subsets. We further restrict our investigation to web server (port 80) traffic. Thus, we are considering only unsolicited SYN/ACK packets to our network from port 80.

Table 1 Data Sets Used in the SYN/ACK Study

Data Set Name	Duration	Number days	Number packets
April	April 4–April 17	14	10,449
May	May 9–May 17	9	23,264
June	June 1–June 15	15	27,845
July	July 1–July 15	15	59,666
Sept	Sept 1–Sept 17	17	210,774
Oct	Sept 19–Oct 15	26	1,253,714
Dec	Oct 28–Dec 12	66	5,421,893
Jan	Jan 1–Jan 31	31	665,392
	Total	193	6,672,997

As can be seen in Tables 1 and 2, we have a much smaller estimate of the number of attacks than the 4000/week of Moore et al. (2001). Part of the reason for this is the fact that we are only looking at one class of attack (SYN flood against a web server). While web servers appear to be the most popular targets, they are not by any means the only ones. There are other attacks, however SYN floods are very popular, due to their effectiveness and the fact that they are hard to defend against, are easy to mount, and are hard to trace back to the originator of the attack.

Figures 2 and 3 depict the data for the eight data sets. In these, the x-axis corresponds to time (in hours) from the start

Table 2 Number of Attacks in each Data Set

Data set	$T = 5\,\text{min}$	$T = 1\,\text{hr}$
April	1510	1231
May	3072	1585
June	2901	2248
July	1727	1220
Sept	3493	1520
Oct	5216	1847
Dec	48,050	3990
Jan	3804	3070
Total	69,773	16,711

of the data set, and the y-axis corresponds to the victim (source) IP address. The IP address is always a 32-bit number with the highest octet in the highest bits. One dot is plotted for every packet (there is considerable overplotting in these pictures, but they serve to illustrate the data).

As can be seen in these figures, there are a number of obvious attacks, as well as some very long-lived attacks. At this resolution it is impossible to count the attacks, and so

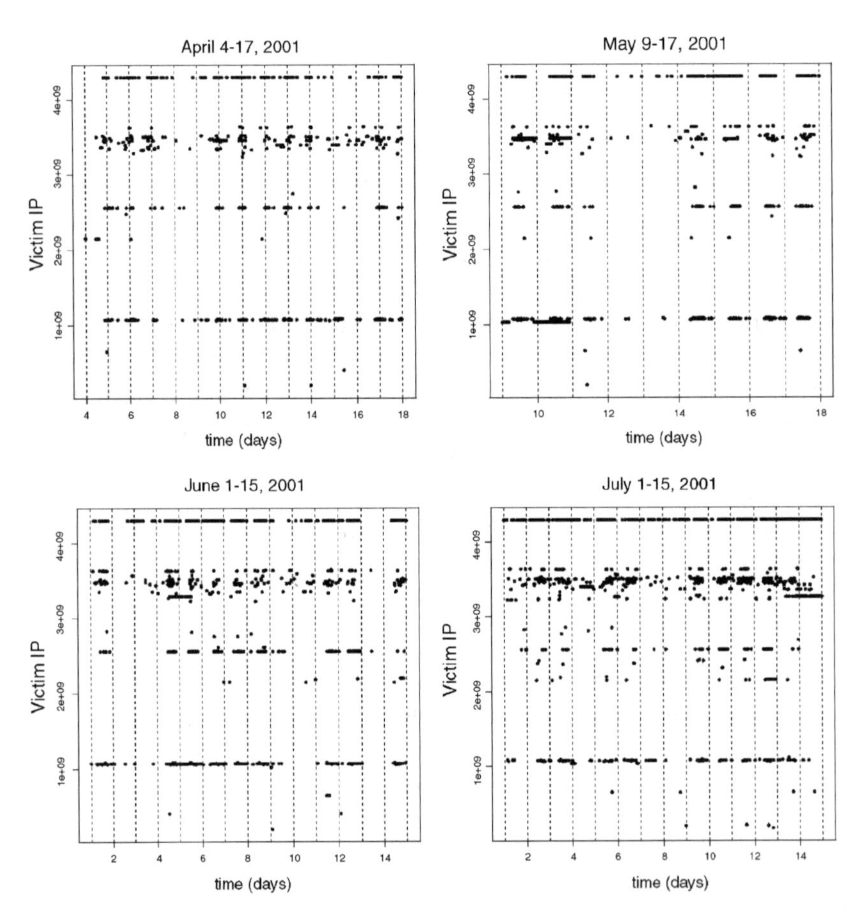

Figure 2 The attacks for the first four data sets. The x-axis is time, the y-axis is a 32-bit number corresponding to victim IP address. A dot is placed for each packet. Days are indicated by dotted lines.

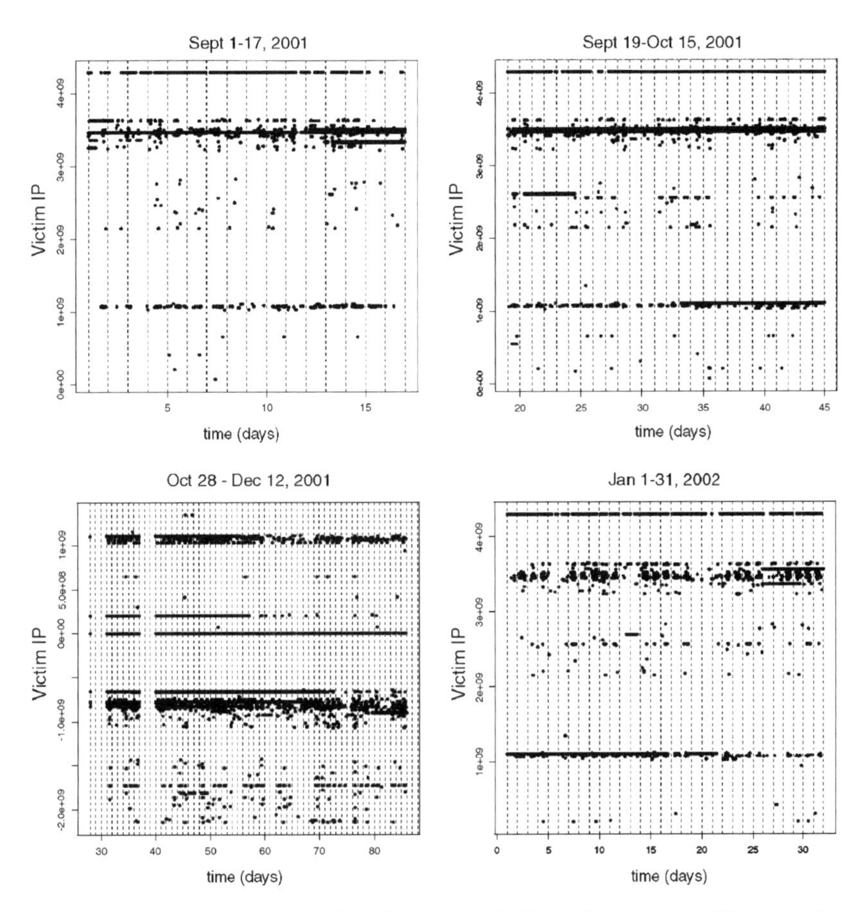

Figure 3 The attacks for the second four data sets. The x-axis is time, the y-axis is a 32-bit number corresponding to victim IP address. A dot is placed for each packet. Days are indicated by dotted lines.

we need to define exactly what we mean by an attack. For our purposes, we define an attack to be a sequence of packets from a single victim such that no gap between packets exceeds a fixed value (T). The results for two values for this threshold are presented in Table 2. If we restrict our definition to those attacks for which we received more than 10 packets, we have the results reported in Table 3.

Table 3 Number of Attacks in Each Data Set for Which There Were More than 10 Packets

Data set	$T = 5\,\text{min}$	$T = 1\,\text{hr}$
April	54	42
May	62	60
June	97	80
July	149	107
Sept	375	192
Oct	1324	177
Dec	6551	414
Jan	263	206
Total	8875	1278

Some care is needed in counting the packets in an attack. Figure 4 depicts the packets from one victim. Note the characteristic "streaking" in this figure. This is a result of resent packets. When the victim does not receive an answer to its SYN/ACK, it waits a small amount of time and then assumes the packet was lost in transit and resends the packet. It repeats this several times, each time increasing the wait period. This results in the "streaks" in the figure, and in an overestimate of the number of attack packets, if this is not taken into account. We define a resent packet to be one which agrees with a previous packet in the source and destination IPs and ports, and the acknowledgment number, and which is received within 1 min of the first such packet. The numbers in Table 3 are computed using this definition, and so resends are not counted in the definition of an attack.

Resends can also be used to help determine whether the packets are backscatter from a denial of service attack, or are a scan of the protected network. One expects to see resends in backscatter. Scan tools that send a single packet per host/port will not show this pattern, while those that send multiple packets will typically not increase the time between packets, nor will they tend to have as large a time between packets as one sees with resent packets.

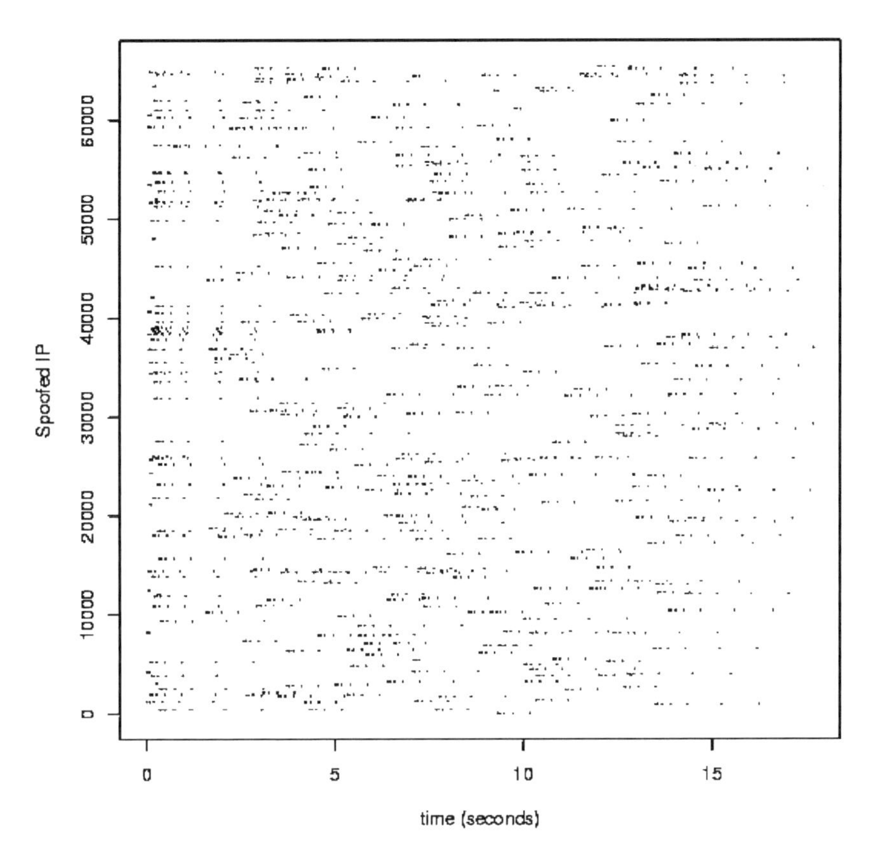

Figure 4 Two thousand one hundred and sixty packets from a single victim computer.

3.2. Attack Statistics

Figures 5 and 6 show histograms for the (log base 2 of the) number of packets detected in the attack, after removing resends, for the different data sets, for the two values of T. Our estimate of the number of packets in the original attack (assuming we believe that the attacker is selecting spoofed IP addresses uniformly, independently, from all 2^{32} possible IP addresses) can be obtained by multiplying the x-axis values by 16. Unfortunately, as we will see, it is not as simple as this. Not all attacks choose randomly from all possible IP

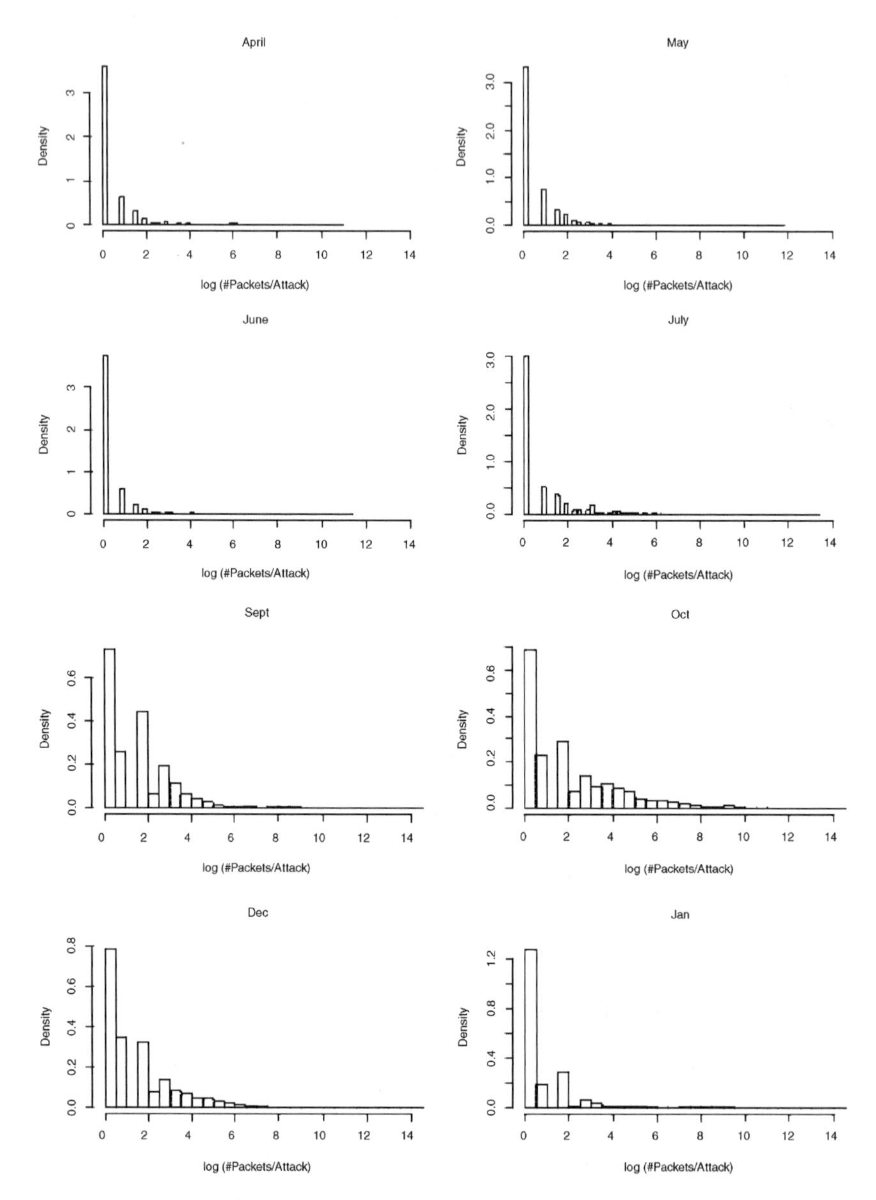

Figure 5 Histogram of the log (base 2) of the number of packets per attack. These counts are computed after the resends have been removed, as described in the text. For these counts, we have used the value $T = 5\,\text{min}$.

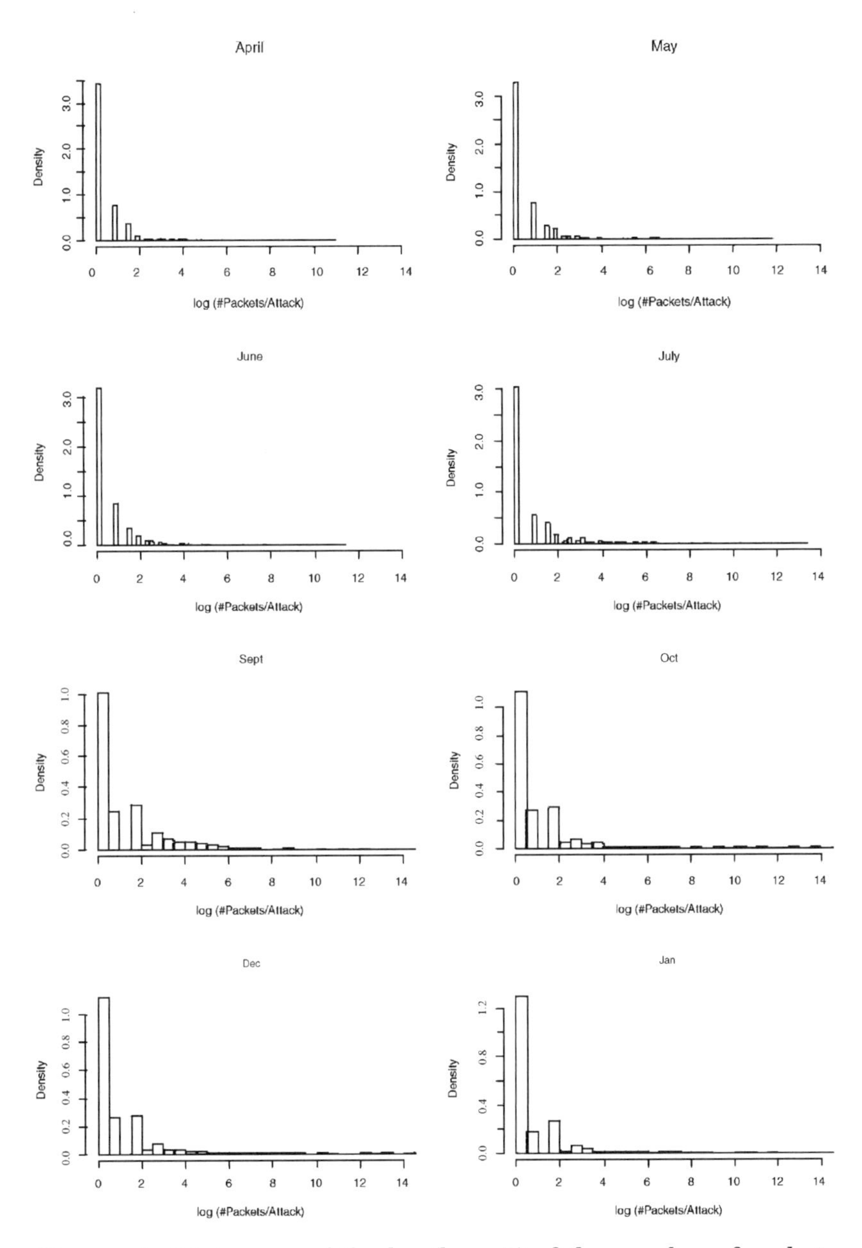

Figure 6 Histogram of the log (base 2) of the number of packets per attack. These counts are computed after the resends have been removed, as described in the text. $T = 1\,\text{hr}$.

addresses, but some do, so these calculations must be done on a case-by-case basis.

One observation is that the densities are surprisingly similar across all the data sets. The histograms appear to support a hypothesis of roughly three modes to the density, indicating (perhaps) the existence of three different types of attacks.

It is likely that many of the packets in the bin at 0 (corresponding to a single packet detected in the attack) represent errors in the process of selecting "unsolicited" packets (for example, dropped SYN packets at the sensor), or are the results of other attacks (such as scans) or errors unrelated to cyber attacks.

Another explanation is that these are low packet rate attacks. Since a SYN flood need only fill the connection table of the victim, and keep it filled, an attack lasting only a few hours need not send more than 2^{16} packets (our estimate of the number of packets in an attack in which we observe 1 packet). Thus, it seems reasonable to suggest that for attacks against single servers (that cannot load-balance using a server farm, for example), attacks of this magnitude might be effective, and popular, accounting for the large number of such "attacks" detected.

We now turn to the question of whether the attack is random, that is, whether the spoofed IP addresses have been (uniformly) randomly selected from all 2^{32} possible IP addresses. Some pictures will be informative. While looking at pictures is subjective, and cannot detect subtle deviations from randomness, it can be very effective in detecting unexpected structure. (Note: in all the analysis which follows we use $T = 5$ min.)

Figures 7 and 8 depict the packets from two victims. In these plots each packet is plotted as a dot, with x value corresponding to time and y value corresponding to the spoofed destination IP address. This is computed from the IP a.b.c.d as 256c + d, since all IP addresses on the protected network agree in the first two octets. Figure 7 seems to pass a "looks random" test, while Figure 8 shows definite nonrandom structure. This manifests itself in two ways. First, it is obvious

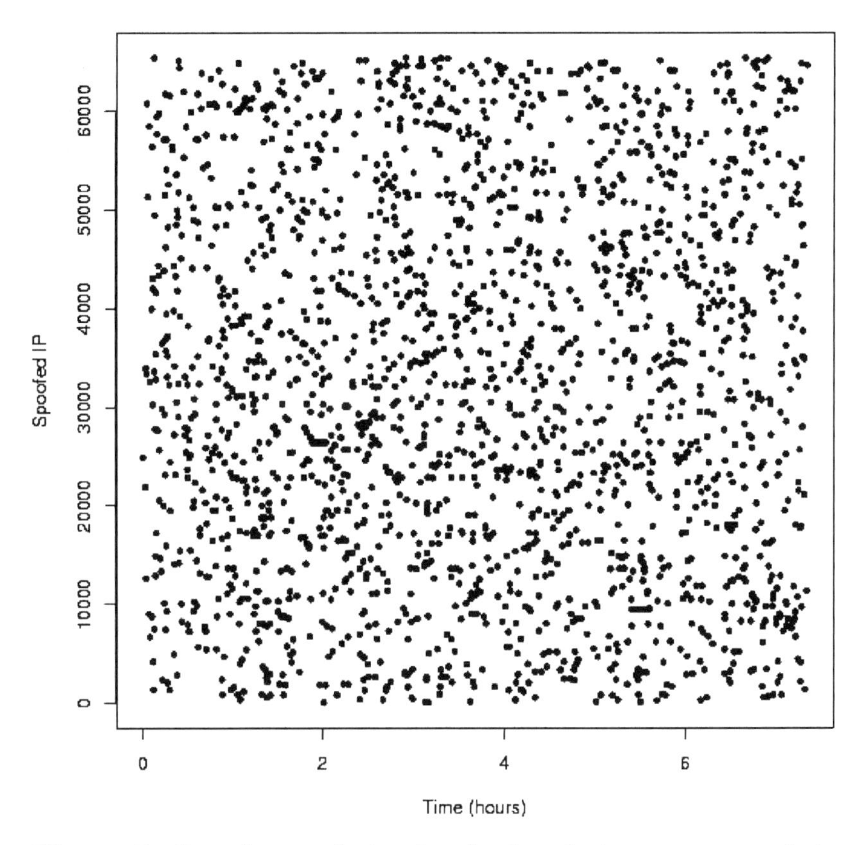

Figure 7 One thousand nine hundred and ninety-seven packets from a single victim computer in April. The x-axis corresponds to the time of arrival of the packet, the y-axis corresponds to the last two octets of the spoofed destination IP address.

that the intensity of the attack is not constant throughout the attack. Second, there is a diagonal structure detectable in the packets, showing a high degree of correlation. This attack does not satisfy our assumptions of independent random selection of spoofed IP addresses. Clearly, care must be taken in the analysis of attacks of this type. One obvious question is how many attacks this figure represents. One could easily argue for one, two or three distinct attacks, depending on the model one used for defining an attack.

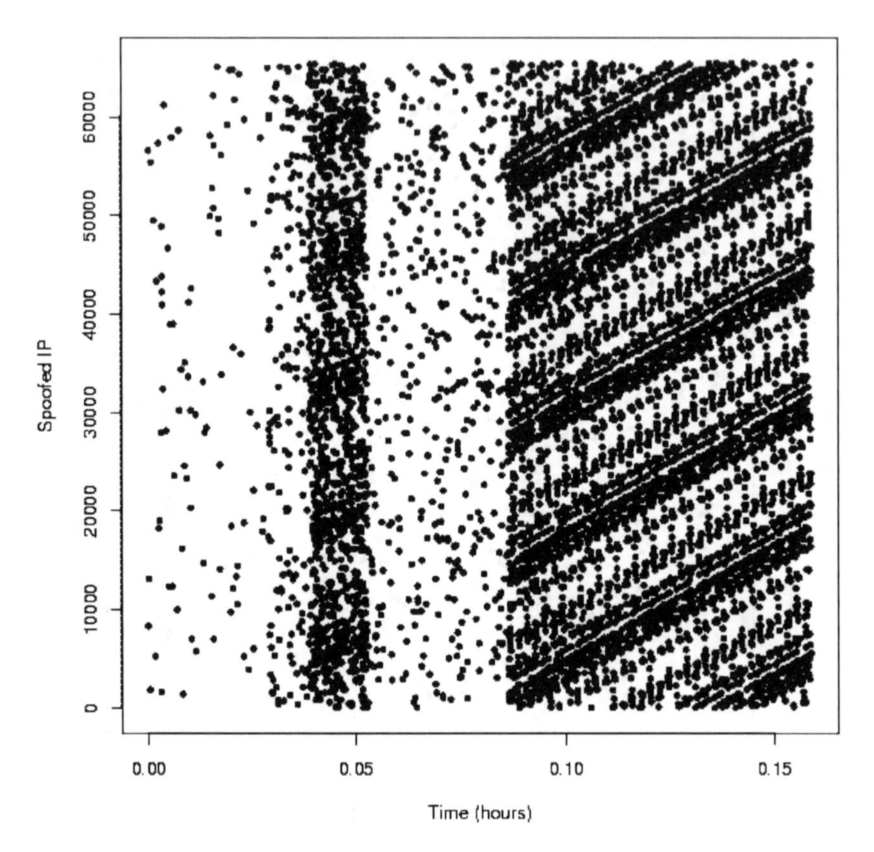

Figure 8 Seven thousand one hundred and thirty-seven packets from a single victim computer in October. The axes are the same as in Figure 7.

Figure 9 depicts attacks against two victims consisting of 9674 and 22,716 packets. These show quite different structure, indicating several different attack tools were used. The top figure shows an attack with linear structure, overlapping an attack that looks to the eye to be fairly random. The bottom figure shows an attack with quite complicated dependence structure, with both a linear component, and some measure of clearly deterministic structure. This latter kind of attack was not observed in the data prior to the October data set.

Because of the systematic nature of the IP address selection in the bottom plot of Figure 9, the data passes a

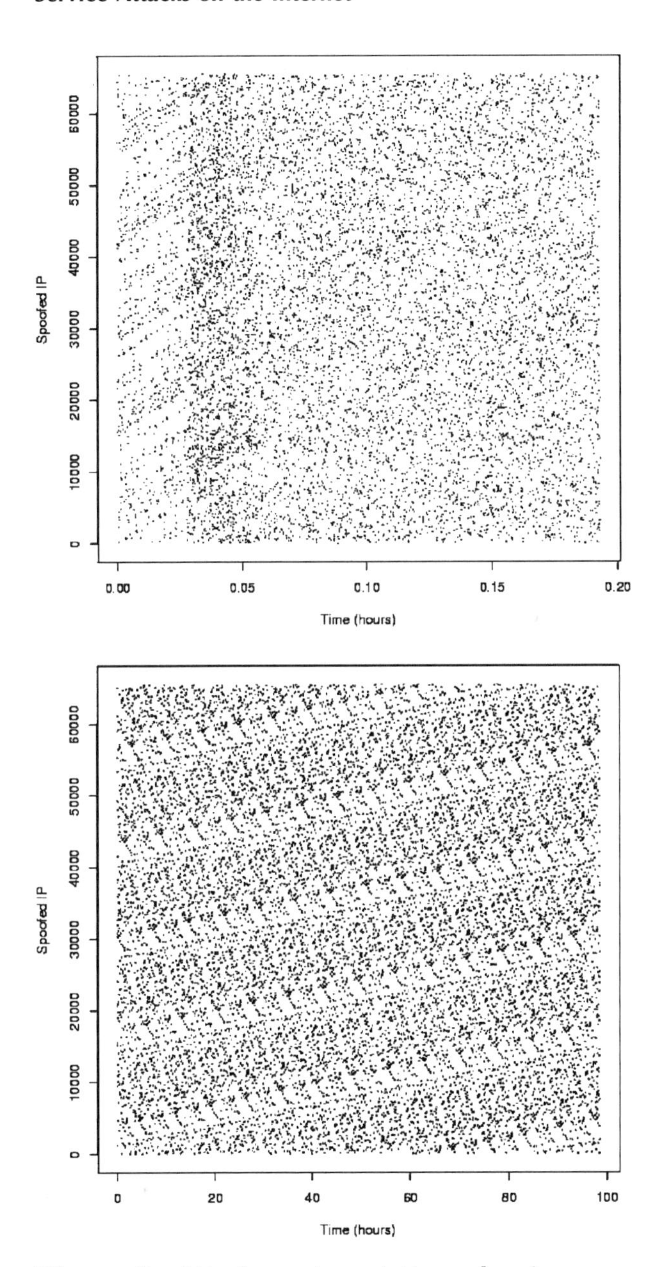

Figure 9 Attacks on two victims, showing nonrandom structure. The top figure represents 9674 packets collected in July, while the bottom represents 22,716 collected in November.

goodness-of-fit test (the Kolmogorov–Smirnov test) with flying colors. This test assumes (and does not test for) independence, and so is invalid for these data.

The above observations indicate that the blind use of goodness-of-fit tests will be of little use for these data. The changing intensity, and structure in many of the attacks make any assessment by a goodness-of-fit test problematic at best. Thus, each attack must be assessed individually, testing the different intensity regions separately. Further, it is vital that tests for dependence be used, in addition to distributional tests.

The number of large attacks (attacks with more than 1000 packets) seems to be increasing in these data. In April, the average was approximately one such attack every two weeks, while by December the rate was approximately two per day. This may be a short time phenomenon (the rate does appear to have dropped to about 1 per day by January), or it may be a result of the increasing availability of attack tools or new attack paradigms. Further data are needed to assess this trend.

It might seem natural to assume that the attacks with linear structure are actually scans of the protected network, rather than backscatter from denial of service attacks. A perusal of the data shows that some of the attacks exhibiting linear structure do not have resent packets associated with them, lending credence to this hypothesis. Of the 69,773 attacks in the $T = 5$ min data, 60,488 contained resent packets. Also, of the 248 attacks consisting of more than 1000 packets (after eliminating resends), 209 of them had resends associated with them. An alternative explanation would be that these victims have been configured to not send retries, but to rather drop the connection if an ACK packet is not received promptly. There is a technique, referred to as "SYN cookies", in which the victim encodes state information in the SYN/ACK packet, and thus does not resend packets. See: http://cr.yp.to/syncookies.html.

The case against the hypothesis that these attacks represent scans of the protected network rests on three observations: first, it is unusual to scan a network from port

80, although one could certainly do this, provided one had the permission necessary to use this port; second, the linear structure does not manifest itself as a sequential pass through the IPs in the domain, but rather, on a small scale, has an apparent random component to it; third the existence of apparent "resend" packets argues against any of the known scan tools. Thus, regardless of the actual nature of the attack, the linear structure still remains to be explained.

Figure 10 is interesting, in that the same pattern is replicated over six different victims, corresponding to addresses xxx.xxx.xxx.3–8. This is an Internet service provider in the

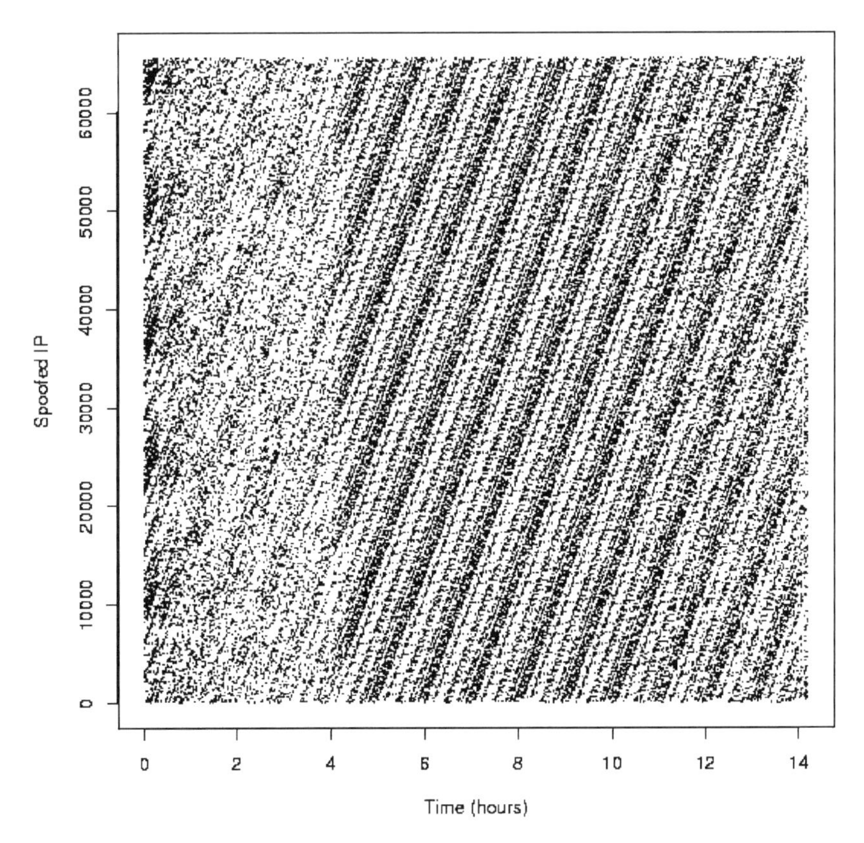

Figure 10 An attack on a United Kingdom Internet Service Provider.

UK, which is obviously using a server farm to load balance. Our hypothesis is that this is a distributed denial of service attack, where different attackers received different IP addresses when the victim IP was resolved through a DNS lookup.

This raises the question of whether the linear pattern that we are seeing is an artifact of the attack or the response of the victim. Perhaps it is the load balancing that is inserting the linear structure into the attack. Perhaps the nonrandom IPs are a result of the timing of responses from the victim, rather than an error in the attack tool's random number generator.

As can be seen in the figure, the character of the attack changes approximately 4 hr into the attack. Is this a change in the packets sent in the attack, or a change in the strategy of the victim(s)? This change occurs approximately simultaneously for all six victims, indicating that in either case the change is coordinated.

Victim action seems unlikely to be the cause, partly from the standpoint that there seems to be little value in it from the point of view of the victim, and partly from further observation of other attacks. A closer look at Figure 9 (top) reveals that there is an overlap between structured and nonstructured attack patterns within the same victim. This is hard to reconcile with the hypothesis that victim response is responsible for the pattern. Thus, we believe that the pattern is a result of the activity of the attacker.

As can be seen in Figure 11, these data are highly correlated, which is hardly surprising given the pictures. One can use this information to build a model of the generating process.

The average interpacket arrival times for the attack in the bottom plot of Figure 9 is 0.0043 with a variance of 1.252×10^{-5}, which is consistent with a single computer sending packets spoofed to come from 2^{16} IP addresses. Note that if one identifies the top and bottom of this figure, the "stripes" in the figure match up. This is further evidence that the attacker is generating IP addresses only from the 2^{16} possible addresses of the protected network. On the other hand, for the top plot in Figure 9, the mean and variance for interpacket arrival times are 2×10^{-5} and 6×10^{-10}. This would seem to

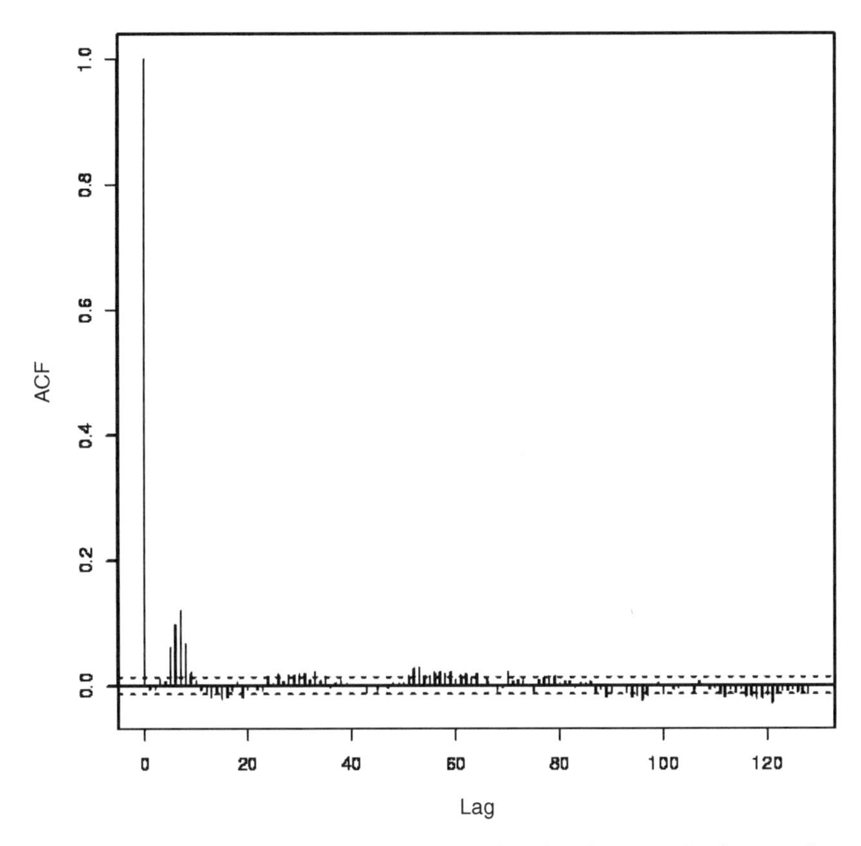

Figure 11 Autocorrelation function for the data in the lower plot of Figure 9, showing statistically significant autocorrelation.

require multiple attackers, and so we hypothesize that this is the result of a distributed denial of service attack.

Figure 12 shows the histograms for the interpacket arrival times for this second attack. Note that the data appear to be approximately lognormal, but seems to have some extra structure. One might hypothesize that this is the result of multiple attackers sending packets which arrive interleaved at the victim, with differences in processor speed, network speed, and route length (and hence delays) causing the result to be a mixture of what one would expect from a single attacker. See, for example, Figure 13, which shows a Q–Q plot for the log of the interpacket arrival times, indicating that

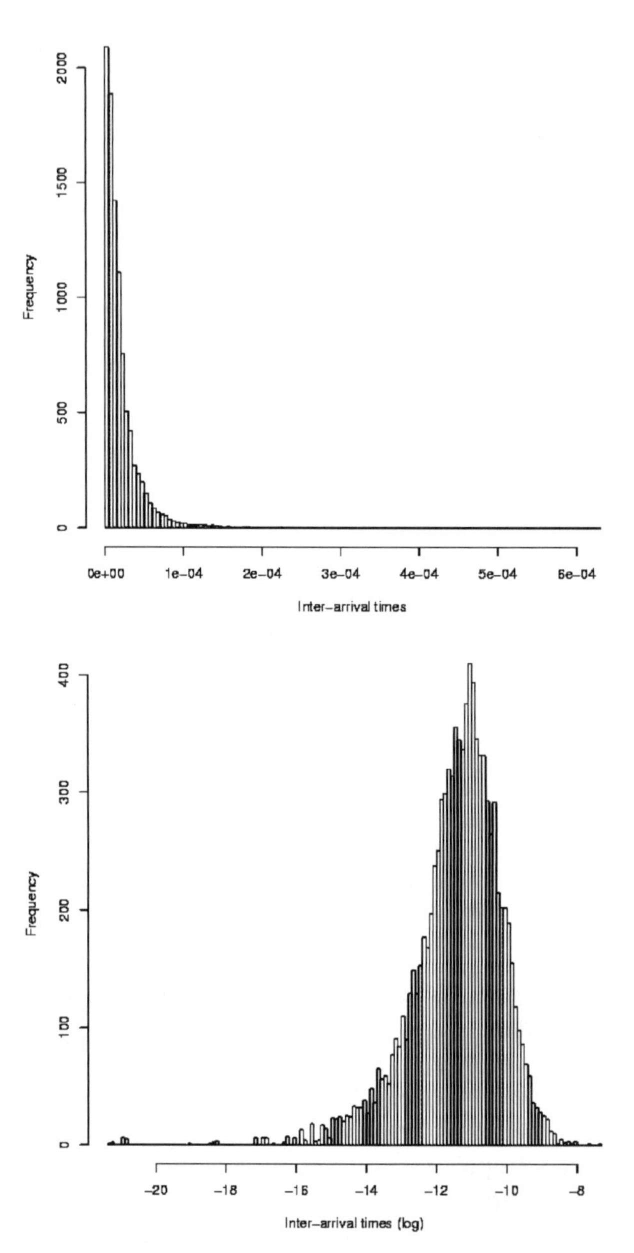

Figure 12 Interpacket arrival times for the attack of Figure 9 (top). The top plot shows a histogram of the interpacket arrival times at the protected network, the bottom shows a histogram of the log of the interpacket arrival times.

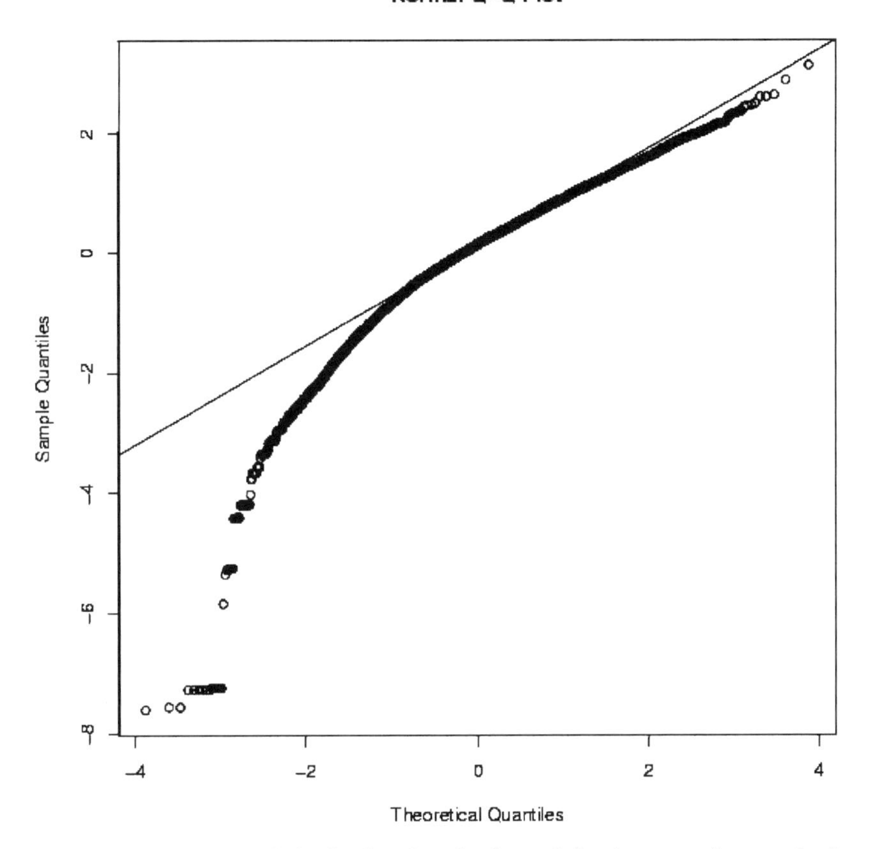

Figure 13 Normal Q–Q plot for the log of the interpacket arrival times for the attack of Figure 9 (top), standardized to have (log) mean 0 and (log) variance 1.

there is a surfeit of low values for the arrival times. This might provide further reason to believe that this is a distributed attack.

Figures 14 and 15 provide a view of the number of attacks ongoing as a function of time. Attacks are defined as packets from individual victims, with no gap between packets of more than 5 mins. For the first four data sets, we see that while attacks occur throughout the time periods considered, there are rarely more than a few attacks at any time, and

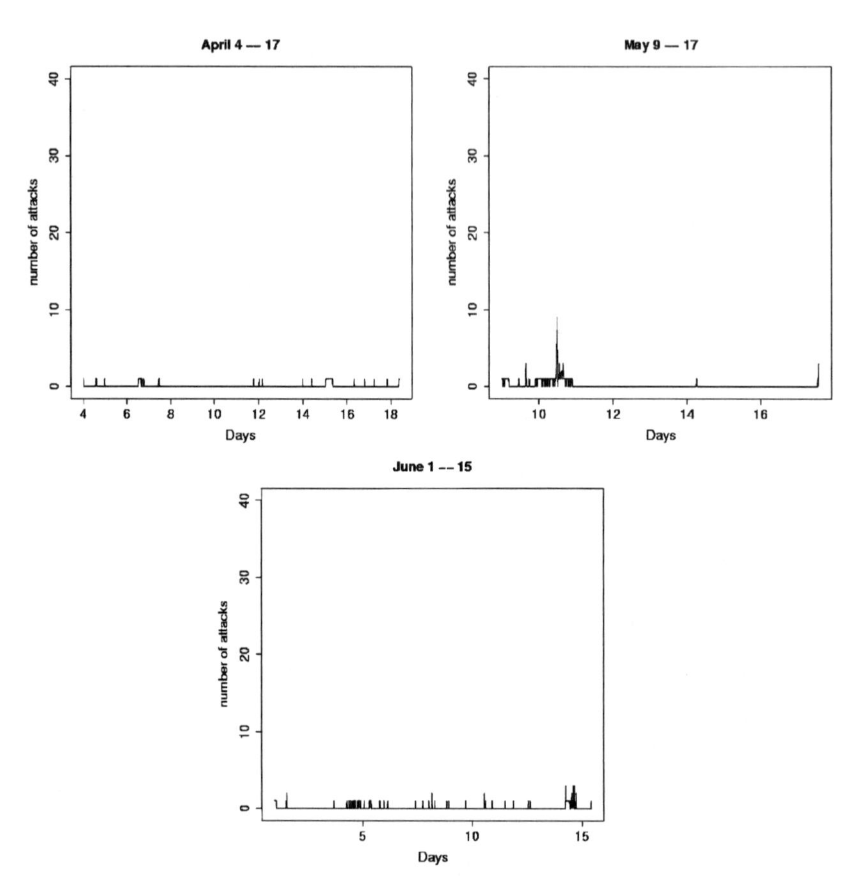

Figure 14 Number of attacks detected as a function of time. Only attacks with more than 10 packets considered.

attacks typically last less than a day. There is some activity between May 10 and May 11, when there were nine simultaneous attacks. Otherwise, the attack level is quite low. The last four data sets show considerable activity. The ramp up in attack levels starts in mid-September, and continues through to late November. At the height of the attacks, there were over 30 victims under attack, and this period of heightened attacks lasted for a month.

These figures were produced using only those attacks consisting of more than 10 packets. Redoing this analysis

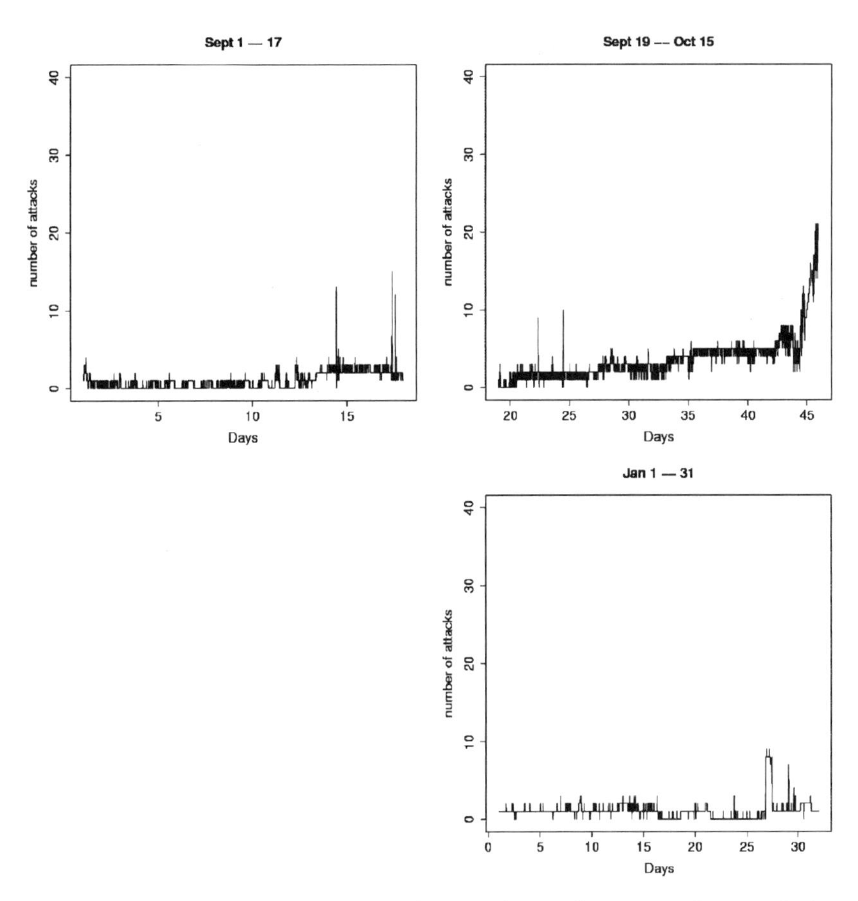

Figure 15 Number of attacks detected as a function of time. Only attacks with more than 10 packets considered. The gaps in the October plot are due to sensor drop out.

including all attacks has a small effect, since small attacks typically have very short duration.

4. CONCLUSIONS

The problem of measuring the number of denial of service attacks on the Internet is a difficult one, since many organizations are hesitant to report these attacks. Even when they do report them, they are often after the fact, and of little value for warning other potential victims of the threat. By utilizing the backscatter packets from certain classes of attacks, we have demonstrated that one can track these attacks in real-time, and we have shown that the attack level on the Internet can be quite high for extended periods of time.

Further work is needed in modeling these attacks, determining the algorithms used for generating packets, and thus providing some ability to classify the attacks. Knowing something about the attack can provide useful information for potential victims to use in defending against the attacks. Also, by monitoring trends in the attacks we can, potentially, identify when new classes of attacks are created, or when a new massive attack is underway.

The problem of determining the impact of the attack on the victim is a difficult one, which we have not addressed here. The victim machine could go down, in which case the backscatter packets would cease, but this may be indistinguishable from a cessation of the attack. It would be of value to determine whether subtle changes in the backscatter packets can be used as indications of the effect of the attack on the victim.

There are some methods available to defend against denial of service attacks, but these are not perfect and have difficulty with large distributed attacks. Incorporating that defense strategy into our analysis might allow us to determine whether the victim is defending against the attack, and thus measure the effectiveness of the defense. With that said, the mere fact that we can track these attacks in real time without the cooperation of the victims and without adding to the load on the network is a powerful and useful tool.

REFERENCES

Chen, E. Y. (2001). AEGIS: an active-network-powered defense mechanism against DDoS attacks. In: Marshall, I. W., Nettles, S., Wakamiya, N. eds. Active Networks: IFIP-TC6 Third International Working Conference, Lecture Notes in Computer Science 2207, Springer, pp. 1–5.

Marchette, D. J. (2001) (2001). *Computer Intrusion Detection and Network Monitoring: A Statistical Viewpoint*. New York: Springer.

Moore, D., Voelker, G. M., Savage, S. (2001). Infering Internet denial-of-service activity. USENIX Security '01. Available on the web at: www.usenix.org/publications/library/proceedings/sec01/moore.html.

Northcutt, S., Novak, J., McLaclan, D. (2001) (2001). *Network Intrusion Detection. An Analyst's Handbook*. Indianapolis, IN: New Riders.

Stevens, W. R. (1994). *TCP/IP Illustrated, Volume 1: The Protocol*. Reading, MA: Addison-Wesley.

13

On the Extinction of the S-I-S Stochastic Logistic Epidemic

RICHARD J. KRYSCIO
Department of Statistics
University of Kentucky,
Lexington, KY, USA

CLAUDE LEFEVRE
Institut de Statistique et
de Recherche Operationelle,
Université Libre de Bruxelles,
Brussels, Belgium

ABSTRACT

We obtain an approximation to the mean time to extinction and to the quasi-stationary distribution for the standard S-I-S epidemic model introduced by [Weiss, G. W., Dishon, J. (1971). On the asymptotic behavior of the stochastic and deterministic models of an epidemic. *Math. Biosci.* 11:261–265.]. These results are a combination and extension of the results of [Norden, R. H. (1982). On the distribution of the time to extinction in the stochastic logistic population model.

Adv. Appl. Prob. 14:687-708.] for the stochastic logistic model, [Oppenheim, I., Shuler, K. E., and Weiss, G. H. (1977). Stochastic theory of nonlinear rate processes with multiple stationary states. *Physica A*, 191–214.] for a model on chemical reactions, [Cavender, J. A. (1978). Quasi-stationary distributions of birth-and-death processes. *Adv. Appl. Prob.* 10:570–586.] for the birth-and-death processes and [Bartholomew, D. J. (1976). Continuous time diffusion models with random duration of interest. *J. Math. Sociol.* 4:187–199.] for social diffusion processes.

1. INTRODUCTION

Consider a sequence of Markov processes $\{I_N(t): t \geq 0\}$ for $N = 2, 3, \ldots,$ where $I_N(t)$ represents the number of infected individuals present at time t in a closed population of N individuals. Note that $N - I_N(t)$ is to be interpreted as the number of susceptible individuals present in the population at time t. For $i = 1, \ldots, N$ and Δt sufficiently small we assume that

$$P\{I_N(t+\Delta t) = j | I_N(t) = i\}. \\ = \begin{cases} (\lambda/N)i(N-i)(\Delta t) + o(\Delta t) & \text{if } j = i+1 \\ \mu i(\Delta t) + o(\Delta t) & \text{if } j = i-1 \\ o(\Delta t) & \text{otherwise.} \end{cases} \tag{1.1}$$

Here λ is the infection rate for susceptibles while μ is the recovery rate for infectives. According to (1.1), the chance of a new infection is proportional to the product of the number of infectives and the number of susceptibles divided by N, the initial number of susceptibles. This process is the stochastic version of the standard S-I-S (susceptible to infective to susceptible) epidemic model introduced by Weiss and Dishon (1971). In its deterministic version, this model has been widely studied and applied to specific diseases (Sanders, 1971; Hethcote et al., 1982). Few results can be found on the stochastic version of this process.

This model is also a special case of a simple birth-and-death process on the integers 0, $1, \ldots, N$ which can be used to describe the propagation of a rumor (Bartholomew, 1982) or to describe a particular chemical reaction (Oppenheim et al., 1977). More recently, this model was discussed at length by Norden (1982) who presented the model as a special case of a stochastic logistic population model. Specifically, Norden used a numerical study to show that when N is large and $\rho = \mu/\lambda < 1$ (i.e., the endemic case), then the mean time to extinction is large; he also suggested an approximation to this mean time based on the quasi-stationary distribution of the process. Oppenheim et al. obtained a crude lower bound for the mean extinction time and in doing so were able to demonstrate analytically that in the endemic case the mean time to extinction increases exponentially at the rate of $(e^{-(1-\rho)}/\rho)^N$. The purpose of this note is to show how to refine this crude lower bound to obtain a reasonable approximation to the mean time to extinction as well as to the quasi-stationary distribution of the process in the endemic case. Some related results applicable to the non-endemic case are also presented.

Before beginning the presentation or our results, we comment briefly on the time-dependent transition probabilities. Note that the Kolmogorov forward equations of the process are given by

$$\begin{aligned} p'_{ij}(t) &= (\lambda/N)(j-1)(N-j+1)p_{i,j-1}(t) \\ &\quad - j(\lambda(N-j)/N + \mu)p_{ij}(t) + \mu(j+1)p_{i,j+1}(t) \qquad (1.2) \end{aligned}$$

where $p_{ij}(t) = P\{I_N(t) = j | I_N(0) = i\}$ for $i = 1, \ldots, N$ and $j = 0, 1, \ldots, N$ and where $p'_{ij}(t) = dp_{ij}(t)/dt$. As pointed out by Norden, a formal solution to the probability generating function of these equations can be based on Huen's system of hypergeometric functions (see Picard, 1965). Since this is an algebraically cumbersome solution, approximate methods of obtaining the time-dependent transition probabilities are discussed. In particular, if we were to assume that $I_N(0) = Ni_0$ where $0 < i_0 < 1$, then we may apply Theorem 3.5 of Kurtz (1971) to show that the sequence of processes $\{Z_N(t) : t \geq 0\}$, where $Z_N(t) = I_N(t)/N$ converges weakly to a Gaussian process having the mean function $i(t)$ and variance function

$$\sigma^2(t) = \frac{1}{2}[\exp(-2(i(t) - i_0)) - 1] + 2\mu \exp\{-2i(t)\} \int_0^t i(v)\mathrm{e}^{-2i(v)}\mathrm{d}v. \qquad (1.3)$$

Here $i(t)$ represents the solution to the deterministic process corresponding to $Z_N(t)$ and satisfies the differential equation

$$i'(t) = \lambda i(t)(1 - i(t)) - \mu i(t) \qquad (1.4)$$

subject to $i(0) = i_0$. This has the "logistic curve" solution

$$i(t) = \begin{cases} \frac{(1-\rho)i_0\mathrm{e}^{(\lambda-\mu)t}}{(1-\rho)+i_0(\mathrm{e}^{(\lambda-\mu)t}-1)} & \text{if } \rho \neq 1 \\ \frac{i_0}{1+i_0\lambda t} & \text{if } \rho = 1. \end{cases} \qquad (1.5)$$

Note that to avoid the absorbing state 0, we must assume that the initial state for $I_N(t)$ increases linearly with N and that this weak convergence result will be applied only over finite time intervals.

2. ON THE MEAN TIME UNTIL EXTINCTION

Let $T_{N,i}$ represent the time until extinction given that $I_N(0) = i$; that is, let

$$T_{N,i} = \inf_t\{I_N(t) = 0\}.$$

In his examination of the stochastic logistic model, Norden (1982) derives the following expression for $\tau_{N,i}$, the mean time until extinction:

$$\tau_{N,i} = \mu^{-1} \sum_{k=1}^{N} \gamma_k \sum_{j=1}^{\min(k,i)} (N - j)!(N\rho)^{N-j+1} \qquad (2.1)$$

where

$$\gamma_k = (N\rho)^{N-k+1}/k(N - k)! \qquad (2.2)$$

(cf. Norden's equation (5.28)). Norden proceeds to derive approximations to $\tau_{N,i}$ for N large and $\rho > 1$. Specifically, by fitting the mean of an exponential distribution to his simulation results, Norden provides the following approximation to the mean extinction time:

$$\tau_{N,i} \approx \exp\{27.9 - 28.3/\rho - 0.28N + 8/\rho^2 + 0.23N/\rho + 0.00001N^2\}. \tag{2.3}$$

This approximation appears only to give the (correct) order of magnitude for $\tau_{N,i}$ when N is large.

In a related paper, Oppenheim et al. (1977) provide a simple lower bound for $\tau_{N,i}$. This lower bound can be derived analytically as follows. First note from (2.1) that $\tau_{N,i}$ is increasing in i. This implies that

$$\tau_{N,i} \geq \tau_{N,1} = \mu^{-1} \sum_{k=1}^{N} \gamma_k (N-1)!(N\rho)^N$$

$$= \rho\mu^{-1} N! \sum_{k=1}^{N} \left\{ (N\rho)^k k(N-k)! \right\}^{-1} \tag{2.4}$$

$$= \rho\mu^{-1} \int_0^{\infty} \mathrm{e}^{-N\rho x} \left\{ (1+x)^N - 1 \right\} / x \, \mathrm{d}x. \tag{2.5}$$

The lower bound suggested by Oppenheim et al. can now be obtained by applying Laplace's approximation to the integral on the right-hand side of (2.5) (see e.g. Pearson, 1983).

$$\tau_{N,i} \geq \frac{\rho}{(1-\rho)\mu} \left(\frac{\mathrm{e}^{-(1-\rho)}}{\rho} \right)^N \sqrt{\frac{2\pi}{N}}. \tag{2.6}$$

This shows that $\tau_{N,i}$ must increase exponentially in N. Also notice that like (2.2), this lower bound does not depend of course upon i. According to Norden's numerical work, this lack of dependence on i is not crucial even for moderately large N whenever $i/N \geq 0.30$. However, this lower bound is much too small to yield even the correct order of magnitude in practice. For example, Table 5 of Norden's simulations show that when $\mu = 0.10$, $\rho = 2/3$, $N = 100$ and $i \geq 30$ we have $\tau_{N,i} = 2.38 \times 10^4$ while the lower bound given by (2.6) yields $\tau_{N,i} \geq 6.80 \times 10^3$.

An improved approximation can, however, be derived as follows. Note that from (2.1)

$$\tau_{N,i} \leq \mu^{-1} \sum_{j=1}^{i} \sum_{k=1}^{N} \gamma_k \int_0^{\infty} \mathrm{e}^{-N\rho x} x^{N-j} \, \mathrm{d}x$$

$$= \sum_{j=1}^{i} B_{N,j} \rho^{j-1} \left\{ \rho \mu^{-1} N! \sum_{k=1}^{N} \left\{ (N\rho)^k k (N-k)! \right\}^{-1} \right\}, \quad (2.7)$$

where $B_{N,j} = (N-j)!N^j/N!$. By applying the approximation used in (2.5)–(2.6) to the quantity inside the brackets of (2.12), we obtain

$$\tau_{N,i} \leq (1-\rho)^{-1} \mu^{-1} \left(\frac{\mathrm{e}^{-(1-\rho)}}{\rho} \right)^N \sqrt{\frac{2\pi}{N}} \sum_{j=1}^{i} B_{N,j} \rho^j. \quad (2.8)$$

To assess the accuracy of this upper bound approximation vs. that suggested by Norden, we refer to Table 1 in which we compare his approximation given by (2.3) to the one presented by (2.8). The exact values were obtained from Norden's Table 6 and assume that i is large and $i/N \geq 0.30$. Clearly (2.8) is an improvement over (2.3) when approximating $\tau_{N,i}$, especially when $N \geq 200$ or $\rho \leq 0.50$.

Table 1 Comparison of (2.3) with (2.8) as approximations to $\tau_{N,i}$ when $\mu = 0.10$, $\rho = 0.50$, 0.66, and N = 100(100)500

		N				
ρ	$\tau_{N,i}$	100	200	300	400	500
0.66	Exact	2.37×10^4	2.10×10^7	2.27×10^{10}	2.63×10^{13}	3.17×10^{16}
	(2.8)	2.20×10^4	2.03×10^7	2.22×10^{10}	2.59×10^{13}	3.14×10^{16}
	(2.3)	2.32×10^4	2.08×10^7	2.28×10^{10}	3.05×10^{13}	5.00×10^{16}
0.50	Exact	1.28×10^9	2.16×10^{17}	4.29×10^{25}	9.09×10^{33}	1.98×10^{42}
	(2.8)	1.25×10^9	2.14×10^{17}	4.26×10^{25}	9.00×10^{33}	1.97×10^{42}
	(2.3)	1.97×10^9	1.74×10^{17}	1.89×10^{25}	2.50×10^{33}	4.03×10^{41}

When $\rho \geq 1$, a different analysis is required. Note that in (2.1) if N is large and k is small, then $(N-j)!/(N-k)!N^k \approx 1$ for all $j = 1, \ldots, k$ so that $\tau_{N,i}$ may be approximated as

$$\tau_{N,i} \approx \mu^{-1} \sum_{k=1}^{N} k^{-1} \sum_{j=1}^{\min(k,i)} \rho^{j-k}$$

which is equivalent to

$$\tau_{N,i} \approx \begin{cases} \rho\mu^{-1}(\rho - 1)^{-1} \sum\limits_{k=1}^{N} (\rho^{\min(k,i)} - 1)/k\rho^k & \text{if } \rho > 1 \\ i\mu^{-1}\left(1 + \sum\limits_{k=i+1}^{N} k^{-1}\right) & \text{if } \rho = 1. \end{cases} \tag{2.9}$$

This shows that when $\rho \geq 1$, the value of $\tau_{N,i}$ depends on i but not upon N provided N is large in comparison to i. In fact, if i is large, we can use the fact that $\{I_N(t) : \tau \geq 0\}$ behaves like a birth-and-death process having the constant birth rate $\lambda(N-i)/N$ and death rate μ. This implies that $T_{N,i}$ converges in distribution to an extreme-valued random variable (see e.g. Barbour, 1975); specifically

$$\left(\mu - \lambda\left(\frac{N-i}{N}\right)\right) T_{N,i} - \log i - \log\left(1 - \frac{\lambda(N-i)}{\mu N}\right) \to W$$

in distribution where W represents the standard extreme value random variate satisfying

$$P(W \leq w) = \exp(-\mathrm{e}^{-w}) \quad \text{for } w > 0.$$

Hence, we find that for $\rho > 1$, N large and i large

$$\tau_{N,i} \approx \left(\mu - \lambda\left(\frac{N-i}{N}\right)\right)^{-1} \left[\gamma + \log i + \log\left(1 - \frac{\lambda(N-i)}{\mu N}\right)\right], \tag{2.10}$$

where γ is Euler's constant. Note that when $(Ni)/N \approx 1$ and when $\rho > 1$, the expression (2.10) is an approximation to

Table 2 Comparison of (2.10) as an approximation to (2.1) (listed in parentheses) when $i = 10$, 25, $\mu = 0.1$, $\rho = 1.25$, 2.0 and $N = 100$, 500, ∞

		N		
ρ	i	100	500	∞
1.25	10	57.4 (72.4)	62.4 (80.0)	63.5 (81.1)
	25	90.1 (94.0)	104.8 (110.9)	109.3 (113.7)
2.00	10	41.5 (45.3)	43.3 (46.4)	45.3 (46.6)
	25	58.2 (59.6)	61.2 (62.4)	62.1 (62.8)

(2.9). Table 2 (in abbreviated form) compares (2.10) with (2.1) for selected values of ρ , N and i. The approximation appears to improve as i increases or as $(\rho-1)$ increases.

3. ON THE QUASI-STATIONARY DISTRIBUTION

For $n = 1, 2\ldots, N$ let

$$q_{i,n}(t) = p_{i,n}(t)/(1 - p_{i,0}(t))$$

represent the conditional probability that there are n infectives in the population at time t given that initially there are i infectives and the disease has not died out at t. Let

$$Q_n = \lim_{t\to\infty} q_{i,n}(t) \quad \text{for } n = 1, 2, \ldots, N.$$

Mandal (1960) showed that $\{Q_n\}$ represents the unique quasi-stationary distribution for the process which is the same regardless of the initial state i. It can be shown that the Q_n satisfy the following system of non-linear equations (see Norden's equations (8.8)–(8.10)):

$$-\mu Q_1^2 = 2\mu Q_2 - (\lambda_1 + \mu)Q_1 \tag{3.1}$$

$$-\mu Q_1 Q_n = \lambda_{n-1}Q_{n-1} - (\lambda_n + n\mu)Q_n + (n+1)\mu Q_{n+1} \quad \text{for } n = 2, \ldots, N-1 \tag{3.2}$$

$$-\mu Q_1 Q_N = \lambda_{N-1} Q_{N-1} - (\lambda_N + N\mu) Q_N, \tag{3.3}$$

where $\lambda_n = \lambda_n\ (N-n)/N$. This system of equations cannot be solved explicitly even for small values of N, and care must be taken when solving these equations on a computer. Cavender (1978, Lemma 1) gives a recursive algorithm which makes it easier to solve these equations numerically on a computer. We now discuss an approximate solution to (3.1)–(3.3).

3.1. The Endemic Case

To begin, consider the reflecting process $\{I^*{}_N(t) : t \geq 0\}$ having the state space $\{1, 2, \ldots, N\}$ and the same infinitesimal transition generator matrix as the process $\{I_N(t) : t \geq 0\}$ when the absorbing state 0 is replaced by a reflecting state at 1. Note that $\{I^*{}_N(t) : t \geq 0\}$ is an ergodic process having a stationary distribution $\{p_n\}$ which satisfies the following system of linear equations:

$$0 = 2\mu p_2 - \lambda_1 p_1 \tag{3.4}$$

$$0 = \lambda_{n-1} p_{n-1} - (\lambda_n + n\mu) p_n + (n+1)\mu p_{n+1} \quad \text{for } n = 2, \ldots, N-1 \tag{3.5}$$

$$0 = \lambda_{N-1} p_{N-1} - (\lambda_N + N\mu) p_N. \tag{3.6}$$

These equations can be solved recursively to yield the well-known solution

$$p_n = \frac{1}{n} \frac{(N-1)!}{(N-n)!} \left(\frac{1}{\rho N}\right)^{n-1} p_1 \quad \text{for } n = 2, \ldots, N, \tag{3.7}$$

where

$$p_1 = \left\{ \sum_{k=1}^{N} \frac{1}{k} \frac{(N-1)!}{(N-k)!} \left(\frac{1}{\rho N}\right)^{k-1} \right\}^{-1}. \tag{3.8}$$

Cavender (1978, Proposition 11) shows that $\{Q_n\}$ is stochastically larger than $\{p_n\}$ in the sense that

$$\sum_{j=1}^{n} Q_j \leq \sum_{j=1}^{n} p_j \quad \text{for } 1,\ldots,N. \tag{3.9}$$

We now show that in the endemic case

$$\sum_{j=1}^{n} Q_j \approx \sum_{j=1}^{n} p_j \quad \text{for } n = 1,\ldots,N. \tag{3.10}$$

To this end, let $\boldsymbol{Q} = (Q_1, \ldots, Q_N)'$ and $P = (p_1, \ldots, p_N)'$; let $\boldsymbol{e}_1$ be the $(N+1) \times 1$ vector $(1,0,\ldots,0)'$ and let F, A and D be $(N+1) \times N$ matrices defined by

$$A = \begin{bmatrix} 1 & 1 & 1 & \cdots & 1 & 1 \\ -\lambda_1 & 2\mu & 0 & \cdots & 0 & 0 \\ \lambda_1 & -(\lambda_2 + 2\mu) & 3\mu & \cdots & 0 & 0 \\ 0 & \lambda_2 & -(\lambda_3 + 3\mu) & \cdots & 0 & 0 \\ \vdots & \vdots & \vdots & . & . & . \\ 0 & 0 & 0 & \cdots & \lambda_{N-1} & -(\lambda_N + N\mu) \end{bmatrix} \tag{3.10}$$

$$\boldsymbol{F} = \begin{bmatrix} \boldsymbol{0} \\ \boldsymbol{e}_1' \\ \boldsymbol{0} \\ \vdots \\ \boldsymbol{0} \end{bmatrix} \quad \text{and} \quad \boldsymbol{D} = \begin{pmatrix} \boldsymbol{0} \\ \boldsymbol{I}_{N\times N} \end{pmatrix}. \tag{3.11}$$

Here 0 is a row vector of zeros and $\boldsymbol{I}_{N\times N}$ is the $N\times N$ identity matrix. Then (3.1)–(3.3) can be written in matrix notation as

$$(\boldsymbol{A} + \mu Q_1 \boldsymbol{D} - \mu \boldsymbol{F})\boldsymbol{Q} = \boldsymbol{e}_1 \tag{3.12}$$

while (3.4)–(3.6) can be written as

$$\boldsymbol{AP} = \boldsymbol{e}_1 \tag{3.13}$$

Note that (3.13) has the familiar solution $\boldsymbol{P} = (\boldsymbol{A}'\boldsymbol{A})^{-1}\boldsymbol{A}'\boldsymbol{e}_1$. Rewrite (3.12) as the perturbed system of equations

$$(\boldsymbol{AQ} - \boldsymbol{e}_1) = y(\boldsymbol{Q}) \tag{3.14}$$

where $y(\boldsymbol{Q}) = -\mu Q_1 \boldsymbol{DQ} + \mu \boldsymbol{FQ}$ can be considered to be a vector of residuals. Notice that by (3.13), $\boldsymbol{P} \approx \boldsymbol{Q}$ if and only if the size of the residual vector $y(\boldsymbol{P})$ is small. But notice that

$$y(\boldsymbol{P})'y(\boldsymbol{P}) = \mu^2(Q_1^2 \boldsymbol{P}'\boldsymbol{D}'\boldsymbol{DP} - 2Q_1 p_1^2 + p_1^2)$$

which by (3.11) is bounded by

$$y(\boldsymbol{P})'y(\boldsymbol{P}) \leq \mu^2(Q_1^2 + \boldsymbol{P}'\boldsymbol{P} + p_1^2).$$

Since P is a vector of probabilities,

$$y(\boldsymbol{P})'y(\boldsymbol{P}) \leq \mu^2(Q_1^2 + p_1^2)$$

which by (3.9) implies

$$y(\boldsymbol{P})'y(\boldsymbol{P}) \leq 2\mu^2 p_1^2.$$

When N is large and $\rho < 1$ we can use (3.8) and (2.6)–(2.8) to show that

$$y(\boldsymbol{P})'y(\boldsymbol{P}) \leq 2\mu^2(1-\rho)^2 N(\rho e^{1-\rho})^{2N}/(2\pi\rho^2).$$

Hence for N large and $\rho < 1$, the size of $y(\boldsymbol{P})'\boldsymbol{y}(\boldsymbol{P})$ will be arbitrarily small, which shows that $\boldsymbol{Q} \approx \boldsymbol{P}$ or that (3.10) holds in the endemic case. We used Cavender's recursive procedure to solve (3.1)–(3.3) exactly and compared this solution to (3.7) and (3.8) by computing

$$d_{N,\rho} = \underset{n=1,\ldots,N}{\text{maximum}} \left| \sum_{i=1}^{n} Q_i - \sum_{i=1}^{n} p_i \right|.$$

We found, for example, that to ensure that $d_{N,\rho} < 0.01$, we must have $N \geq 30$ when $\rho = 0.5$ but $N \geq 150$ when $\rho = 0.75$. Hence, the use of P for Q depends on both size of N and $1-\rho$.

Clearly P is much easier to work with than Q. Oppenheim et al. show that the stationary distribution P is very peaked about the point $\boldsymbol{N}(1-\rho)$ and that the reflection process is with high probability likely to be in this state whenever $\rho < 1$. Since $\boldsymbol{P} \approx \boldsymbol{Q}$, then the same statement is true about the original absorbing process. Hence, in the endemic case, the stochastic process behaves like the deterministic process.

We note that this approximation fails when $\rho \geq 1$. In the case $\rho > 1$, it is easy to show that $Q_1 > Q_2 > \cdots$ and that $p_1 > p_2 > \cdots$. In fact, both probability distributions are very peaked at the state 1, but, unfortunately, $Q_1 - p_1$ can be substantial. For example, when, $N = 100$ and $\rho = 2.0$ we have $Q_1 = 0.514$ while $p_1 = 0.726$.

3.2 The Non-endemic Case

To obtain an approximation to P which is valid for $\rho > 1$, consider the birth-and-death process having the state space $\{1, 2, \ldots, N\}$ and the same infinitesimal generators as the process $\{I_N(t) : t \geq 0\}$ except that the process has one infective who can never recover. If $\{m_n\}$ denotes the stationary distribution of this "modified" process, then this distribution must satisfy the following system of equations:

$$\begin{aligned}
0 &= \mu m_2 - \lambda_1 m_1 \\
0 &= \lambda_{n-1} m_{n-1} + n\mu m_{n+1} - (\lambda_n + (n-1)\mu) m_n \\
&\quad \text{for } n = 2, \ldots, N-1 \\
0 &= \lambda_{N-1} m_{N-1} - (\lambda_N + (N-1)\mu) m_N.
\end{aligned}$$

These equations can be solved recursively to yield

$$m_n = \frac{(N-1)!}{(N-n)!} (\rho N)^{-(n-1)} m_1 \quad \text{for } n = 2, \ldots, N \tag{3.15}$$

where

$$m_1 = \left[\sum_{k=1}^{N} \frac{(N-1)!}{(N-k)!} (\rho N)^{-(k-1)} \right]^{-1}. \tag{3.16}$$

The definition of this modified process was motivated by a stochastic model for the spread of rumors due to Bartholomew

(1976), cf. his case $w=1$. Specifically, Bartholomew considers a birth and death process in which the birth rate parameters λ_0, $\lambda_1,\ldots,$ of the process $\{I_N(t):t\geq 0\}$ are replaced by $\tilde{\lambda}_0,\tilde{\lambda}_1,\ldots,$ respectively, where $\tilde{\lambda}_n=\lambda(n+1)(N-n)/N=\lambda_n+\lambda(N-n)/N$, while the death rate parameters remain unchanged. Denoting this new process by $\{\tilde{I}_N(t):t\geq 0\}$, it is clear that while state 0 can be reached in this process, it is not an absorbing state. Let $\pi_n=P\{\tilde{I}_N(\infty)=n\}$ for $n=0,\ 1,\ldots,\ N$ and $p^*{}_n=\pi_n/(1-\pi_0)$ for $n=1,\ldots,\ N$. The probability distribution of the number of infectives given that the disease is really present in the population satisfies the following relationship for $n=1,\ldots,\ N-1$:

$$\frac{p_n^*}{p_{n+1}^*}=\frac{\pi_n}{\pi_{n+1}}=\frac{(n+1)\mu}{\tilde{\lambda}_n}=\frac{\rho N}{(N-n)}.$$

Using (3.15) and (3.16), we find that

$$\frac{m_n}{m_{n+1}}=\frac{\rho N}{(N-n)}. \tag{3.17}$$

This shows that $m_n=p^*{}_n$ for $n=1,\ldots,N$. As first pointed out by Bartholomew, this is a surprising result: the probability distribution of the process with one external infective given the disease is present is equivalent to the probability distribution of the process with one permanent infective.

Of current interest is the following inequality obtained from (3.17):

$$\frac{m_n}{m_{n+1}}<\frac{\rho N(n+1)}{n(N-n)}=\frac{p_n}{p_{n+1}}.$$

This means that $\{m_n\}$ is larger than $\{p_n\}$ in the sense of likelihood ratio (see e.g., Ross, 1983). By a well-known result, we deduce that $\{m_n\}$ is stochastically larger than $\{p_n\}$, i.e.,

$$\sum_{j=1}^{n}p_j\geq\sum_{j=1}^{n}m_j \quad \text{for } n=1,\ldots,N \tag{3.18}$$

We now conjecture that

$$\sum_{j=1}^{n} m_j \leq \sum_{j=1}^{n} Q_j \leq \sum_{j=1}^{n} p_j \quad \text{for } n = 1,\ldots,N$$

part of which is established by (3.9) and (3.18). This missing component, namely $\sum_{j=1}^{n} m_j \leq \sum_{j=1}^{n} Q_j$, appears to be true from our numerical work. In fact, if $M = (m_1, \ldots, m_N)'$, we found that $\boldsymbol{M} \approx \boldsymbol{Q}$ in the non-endemic case whenever $\rho - 1$ is large. To be more specific, let

$$d^*_{N,\rho} = \underset{n=1,\ldots,N}{\text{maximum}} \left| \sum_{i=1}^{n} Q_i - \sum_{i=1}^{n} m_i \right|,$$

we found that to ensure that $d^*{}_{\mathrm{N},\rho} < 0.01$, we needed to have $N \geq 40$ and $\rho \geq 2$. It remains to establish a reasonable approximation for $|\rho - 1|$ small.

REFERENCES

Barbour, A. D. (1975). The duration of the closed stochastic epidemic. *Biometrika* 62:477–482.

Bartholomew, D. J. (1976). Continuous time diffusion models with random duration of interest. *J. Math. Sociol.* 4:187–199.

Bartholomew, D. J. (1982). *Stochastic Models for Social Processes*. New York: Wiley.

Cavender, J. A. (1978). Quasi-stationary distributions of birth-and-death processes. *Adv. Appl. Prob.* 10:570–586.

Hethcote, H. W., Yorke, J. A., and Nold, A. (1982). Gonorrhea modeling: a comparison of control methods. *Math. Biosci.* 58: 93–109.

Kurtz, T. G. (1971). Limit theorems for sequences of jump Markov processes approximating ordinary differential processes. *J. Appl. Prob.* 8:344–356.

Mandl, P. (1960). On the asymptotic behavior of probabilities within classes of states of a homogeneous Markov process (in Russian). *Časopis Pěst. Mat.* 85:448–456.

Norden, R. H. (1982). On the distribution of the time to extinction in the stochastic logistic population model. *Adv. Appl. Prob.* 14:687–708.

Oppenheim, I., Shuler, K. E., and Weiss, G. H. (1977). Stochastic theory of nonlinear rate processes with multiple stationary states. *Physica A* 88(2):191–214.

Picard, P. (1965). Sur les modèles stochastiques logistiques en démographie. *Ann. Inst. H. Poincaré B* 2:151–172.

Pearson, C. E. (1983). *Handbook of Applied Mathematics*. Princeton, NJ: Van-Nostrand-Reinhold.

Ross, S. (1983). *Stochastic Processes*. New York: Wiley.

Sanders, J. L. (1971). Quantitative guidelines for communicable disease control programs. *Biometrics* 27:883–893.

Weiss, G. W., and Dishon, J. (1971). On the asymptotic behavior of the stochastic and deterministic models of an epidemic. *Math. Biosci.* 11:261–265.

14

Secure Introduction of One-way Functions

DENNIS VOLPANO

Computer Science Department, Naval Postgraduate School, Monterey, CA, USA

ABSTRACT

Conditions are given under which a one-way function can be used safely in a programming language. The security proof involves showing that secrets cannot be leaked easily by any program meeting the conditions unless breaking the one-way function is easy. The result is applied to a password system where passwords are stored in a public file as images under a one-way function.

1. INTRODUCTION

One-way functions play an important role in security. Roughly speaking, a function f is one-way if for all w, it is easy to compute $f(w)$ but hard to find a z, given $f(w)$, such that $f(z)=f(w)$. One-way functions come in different flavors. Some are permutations, while others are hash functions. They operate upon an arbitrary-length preimage message, producing what is called a message digest. A message digest may have fixed length. Examples of hash functions include, MD5, which produces a 128-bit digest, and SHA1, which yields a 160-bit digest (Schneier, 1996). The hardness property coupled with fixed-length digests make certain one-way hash functions appealing for storing passwords on systems and creating preimages of digital signatures. The main result of this paper is independent of the flavors of one-way functions.

A related property is claw-freeness (Rivest, 1990). A hash function f is said to be claw-free if it is hard to find a pair (x, y), where $x \neq y$, such that $f(x)=f(y)$. For a small message space, a hash function may be one-way but fail to be claw-free due to a birthday attack. The basic idea is that one can significantly reduce the size of a message space and still expect to find, with reasonable probability, two messages that collide. Whether this is an issue depends on the application. This paper is not concerned with the claw-free property.

In this paper, we are interested in identifying conditions under which a one-way function can be used in a programming language safely and with more flexibility than what an information-flow property like noninterference (Volpano et al., 1996) allows. For instance, a cryptographic API for a programming language might include MD5. In this case, the conditions should make leaking a secret using MD5 in any program as hard as inverting MD5. This is a security property under which we justify downgrading MD5 message digests.

We start with the definition of one-way functions from Sipser (1997). A function $f : \Sigma^* \rightarrow \Sigma^*$ is one-way if

1. $|w| = |f(w)|$ for all w (f is length preserving),
2. f is computable in polynomial time, and
3. for every probabilistic polynomial time Turing machine M, every k, and sufficiently large n, if we pick a random w of length n and run M on input $f(w)$,

$$\Pr[M(f(w)) = y \quad \text{where} f(y) = f(w)] \leq n^{-k}.$$

The first and second conditions are irrelevant as far as our main result is concerned. The probability in the third condition is taken over the random choices made by M and the random choice of w. The third condition effectively merges two properties that we need to distinguish for the purpose of constructing a security proof. One is simply the likelihood that f avoids collisions with respect to a given input distribution. This property we term collision resistance. The other is purely a property about inversion where the third condition becomes $\Pr[M(f(w)) = w] \leq n^{-k}$. This is the one-wayness property of f.

If string w is considered private (high) then we might argue that $f(w)$ could be considered public (low) based on the one-wayness of f. However, it is actually unsound to do so unless care is taken in what we allow as arguments to f. For instance, suppose f is a one-way function, variable h stores a k-bit password, and mask is a low variable. Then consider the code in Figure 1.

It copies (leaks) h to low variable l in time linear in k. (It might fail to copy every bit of h because of collisions, but this may be unlikely depending on the collision resistance of f.)

```
l := 0;
mask := 2^(k - 1);
while mask ≠ 0 do
  if  f (h) = f (h |
mask) then
    l := l | mask;
  mask := mask >> 1
```

Figure 1 An efficient leak of h.

However, there are practical examples of where we need to treat a message digest as low. Consider a challenge–response protocol. A participant may respond publicly with a message digest computed over a shared secret and a public challenge it receives. We want the digest to be treated as low. Another example is password checking. If h stores a password, then a simple password checker is given by the assignment

$$b := (f\,(h) = f\,(r))$$

where b is a low output variable and r is the input to the checker. We would expect r and h to be high variables. After all, r may match h, and indeed usually will. However, the result of comparing the message digests must be low.

So we want a set of conditions for a programming language that prohibits abuses of one-way functions, as in Figure 1, yet recognizes legitimate downgrading by them in other situations. This paper describes such a set of conditions via a type system. Further, we need a sense in which these conditions are sound. They are certainly not sound with respect to noninterference Volpano et al. (1996) due to downgrading. However, they are sound in the following sense. It can be proved that leaking the secret contents of a variable h using any program P that meets the conditions is as hard as learning h with a program where access to h is prohibited, but the program can access $f(h)$, call f on inputs of its choice and flip a coin. And deducing h in this context clearly amounts to inverting $f(h)$ using a probabilistic Turing machine. By the one-wayness of f then, we expect P to succeed with very low probability in polynomial time, for sufficiently long and uniformly distributed values of h.

Informally, we reduce the problem of inverting a one-way function to that of leaking a secret h via a well-typed program. We begin with a well-typed program that can access h directly and show that its low computation can be simulated by a program with no references to high variables except in calls of the form $f(h)$ and in comparisons of the form $f(h) = f(r)$, for a high read-only variable r. The latter comparisons are then eliminated by an independent random variable whose distribution is governed by the collision resistance of f with respect to the well-typed program's input distribution. (It is irrelevant that we may not know the distribution because the reduction only relies upon its existence.) The result is a program that uses f, $f(h)$ and an independent random variable to simulate the well-typed program's low computation with at least the same probability of success and with at most a constant increase in time complexity. Therefore, any bound on the probability of finding h from $f(h)$ within polynomial time can apply to the probability of leaking h with a well-typed polynomial-time command. This is a security property that applies, for instance, to the simple password checker above.

2. THE LANGUAGE AND SEMANTICS

A program is expressed in an imperative language:

$$\begin{aligned}(\text{expr})\, e ::= {} & x \mid h \mid n \mid f(e) \mid f(h) = f(r) \mid e_1 + e_2 \mid e_1 < e_2 \\ & \mid e_1 = e_2 \mid e_1 \,\&\, e_2 \mid e_1 \gg e_2 \mid (e_1 \mid e_2) \\ (\text{cmd})\, c ::= {} & \text{skip} \mid x := e \mid c_1; c_2 \mid \text{if}\, e\, \text{then}\, c_1\, \text{else}\, c_2 \mid \text{while} \\ & e\, \text{do}\, c\end{aligned}$$

Metavariable x ranges over identifiers that are mapped by memories to integers, n ranges over integer literals, f is a function mapping integers to integers, and h and r are read-only variables. There are three bitwise operators ($\&$, $\gg$, $|$). Integers are the only values; we use 0 for false and nonzero for true.

A standard transition semantics for the language is given in Figure 2. It is completely deterministic and defines a transition function $\rightarrow$ on configurations.

A memory μ is a mapping from variables to integers. A configuration is either a pair (c, μ) or simply a memory μ. In the first case, c is the command yet to be executed; in the second case, the command has terminated, yielding final memory μ. As usual, we define $\kappa \xrightarrow{0} \kappa$, for any configuration κ, and $\kappa \xrightarrow{k} \kappa''$, where $\kappa > 0$, if there is a configuration κ' such that $\kappa \xrightarrow{k-1} \kappa'$ and $\kappa' \rightarrow \kappa''$.

Expressions are evaluated atomically and we extend the application of μ to expressions, writing $\mu(e)$ to denote the

(NO-OP) $(\texttt{skip}, \mu) \rightarrow \mu$

(UPDATE)
$$\frac{x \in \mathrm{dom}(\mu)}{(x := e, \mu) \rightarrow \mu[x := \mu(e)]}$$

(SEQUENCE)
$$\frac{(c_1, \mu) \rightarrow \mu'}{(c_1; c_2, \mu) \rightarrow (c_2, \mu')}$$

$$\frac{(c_1, \mu) \rightarrow (c'_1, \mu')}{(c_1; c_2, \mu) \rightarrow (c'_1; c_2, \mu')}$$

Figure 2 Transition semantics.

value of expression e in memory μ. We say that $\mu(f(e)) = f(\mu(e))$, $\mu(e_1 + e_2) = \mu(e_1) + \mu(e_2)$, and so on. The other expressions are handled similarly. Note that $\mu(e)$ is defined for all e, as long as every identifier in e is in dom(μ).

2.1. Probabilistic Execution

In our reduction of Section 4, we talk about the probabilistic simulation of a command with respect to a joint distribution d for its free variables (d is finite for a given command if memories are mappings to k-bit integers for a fixed k). A simulation may need to flip a coin but this occurs only once,

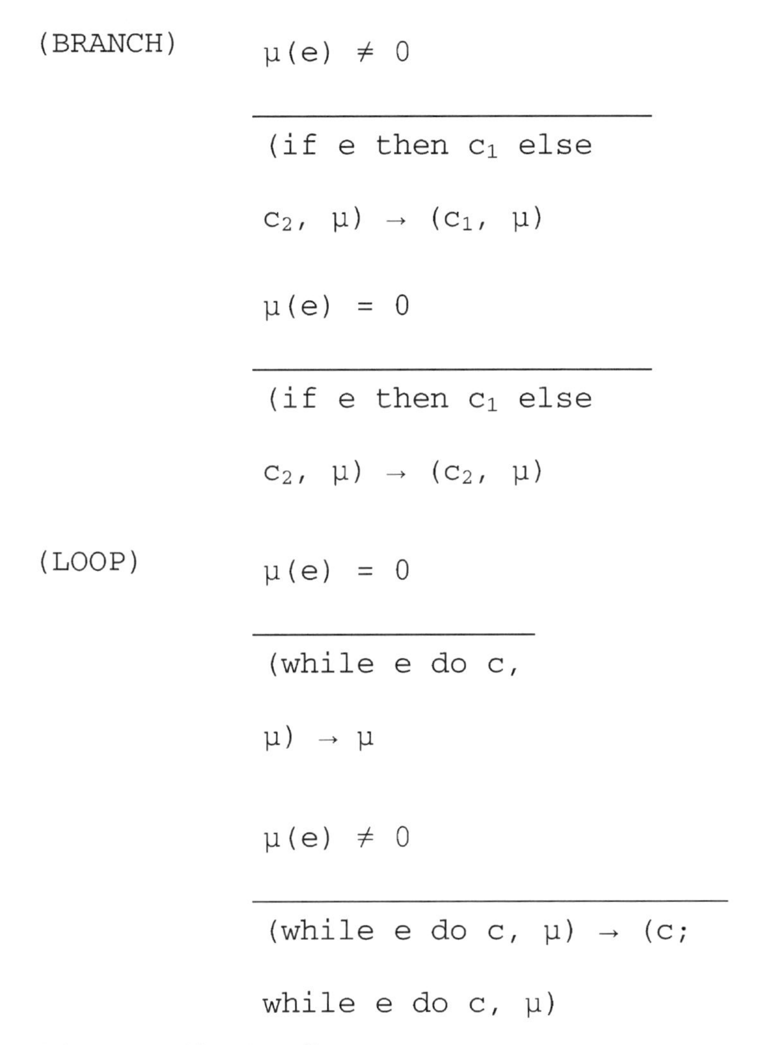

Figure 2 (Continued).

at the start of an execution, and therefore we can achieve the effect by introducing an independent random variable as input to the simulation. Although this keeps the simulation deterministic, it still calls for attention in the semantics because the input variable must be initialized from a probability space prior to execution, apart from other free variables (Kozen, 1981).

The free and random variables of a command can be treated uniformly as just free variables if a command is represented as a discrete Markov chain, the states of which are configurations (Volpano and Smith, 1999). The idea is to execute a command simultaneously in all memories that map exactly its free variables. For each such memory μ, it begins execution in μ with probability $d(\mu)$. An execution then becomes a sequence of probability measures on configurations. The stochastic matrix T of the Markov chain in this case is trivial; each row of the matrix is a point mass. That means there is no splitting of mass after execution begins, only accumulation of it (cf. p. 337 of Kozen, 1981). Each measure in a sequence is determined by taking the linear transformation of the immediately preceding measure with respect to T. See Volpano and Smith (1999) for details.

For example, execution of $y := \neg x$ is given in Figure 3 relative to a particular joint distribution for the four possible memories.

We say that $y := \neg x$ terminates in memory $[x := 0, y := 1]$ in one step with unconditional probability 5/8 and in $[x := 1, y := 0]$ in one step with probability 3/8.

As another example, consider the loop

$$\text{while}\, x \,\text{do}\, x := \neg x$$

```
(y := ¬ x, [x := 0, y := 0]) : 3/8
(y := ¬ x, [x := 0, y := 1]) : 1/4
(y := ¬ x, [x := 1, y := 0]) : 1/8
(y := ¬ x, [x := 1, y := 1]) : 1/4

              ↓

[x := 0, y := 1] : 3/8 + 1/4 = 5/8
[x := 1, y := 0] : 1/8 + 1/4 = 3/8
```

Figure 3 Execution of $y := \neg x$ as a sequence of measures.

whose execution is given in Figure 4 for a particular distribution. Mass accumulates at $[x := 0]$ (or in the notation of Volpano and Smith, 1999, at $(\{\ \}, [x := 0]))$ in the final step. We say the command terminates in $[x := 0]$ in three steps with unconditional probability 1.

In general, suppose μ is a memory, c has free variables $x_1, \ldots, x_n$ and $\mu_1, \ldots, \mu_m$ are memories with domain $x_1, \ldots, x_n$ from which c terminates in μ in at most k steps. If d is a joint distribution for $x_1, \ldots, x_n$, then we say c terminates in μ in k steps with unconditional probability $d(\mu_1) + \square\square\square + d(\mu_m)$.

3. THE TYPE SYSTEM

Following previous work, the types are as follows:

$$\begin{array}{ll} \text{(data types)} & \tau ::= L \mid H \\ \text{(phrase types)} & \rho ::= \tau \mid \tau\,\text{var} \mid \tau\,\text{cmd} \end{array}$$

```
(while x do x := ¬ x, [x := 0]) : 1/3
(while x do x := ¬ x, [x := 1]) : 2/3

↓

[x := 0]) : 1/3
(x := ¬ x ; while x do x := ¬ x, [x := 1]) : 2/3

↓

[x := 0]) : 1/3
(while x do x := ¬ x, [x := 0]) : 2/3

↓

[x := 0]) : 1/3 + 2/3 = 1
```

Figure 4 Execution of while x do $x := \neg x$ as a sequence of measures.

The data types are just the security levels low and high. The rules of the type system are given in Figure 5.

Here, γ is a typing that maps variables (perhaps read-only) to types of the form τ var or τ. If $\gamma(x) = \tau$, then we say that x is a read-only variable in γ. We distinguish h and r as special read-only variables in that

$$\gamma(h) = H = \gamma(r)$$

for all typings γ.

The typing rules for the other binary operators are similar to that for EQ. Notice that where downgrading is taking place, specifically in rules QUERY and IMAGE, it is done with respect to read-only variables, namely r and h. This is key to getting a reduction. It is these two rules that break traditional noninterference. Rule IMAGE is useful when typing the code of a challenge–response protocol, for instance, the GNU implementation of CHAP. It allows a low digest to be computed over a challenge and a secret (the challenge response), the concatenation of which is the value of h. Rule QUERY is useful in password-checking contexts. More is said about these applications in Sec. 5.

Notice that the code in Figure 1 is not well typed. Expression $f(h \mid \text{mask})$ can only be typed using rule HASH, forcing it to have type H since $\gamma(h) = H$ for all γ. But then the guard of the conditional has type H, while its body has type L cmd which cannot be reconciled.

4. THE REDUCTION

The basic idea is to show that every well-typed command's low computation can be simulated, with at most a constant increase in time complexity, by a command whose only references to high variables are in calls to f. However, we are not finished. The simulation still contains calls of the form $f(h)$ and $f(r)$. Instances of $f(h)$ can remain because they form the input to a command (the adversary) for computing h, but all calls $f(r)$ must be eliminated.

(INT) $\gamma \vdash n : L$

(IMAGE) $\gamma \vdash f(h) : L$

(QUERY) $\gamma \vdash f(h) = f(r) : L$

(CONST) $$\frac{\gamma(x) = \tau}{\gamma \vdash x : \tau}$$

(R-VAL) $$\frac{\gamma(x) = \tau \text{ var}}{\gamma \vdash x : \tau}$$

(EQ) $$\frac{\gamma \vdash e_1 : \tau, \quad \gamma \vdash e_2 : \tau}{\gamma \vdash e_1 = e_2 : \tau}$$

(HASH) $$\frac{\gamma \vdash e : \tau}{\gamma \vdash f(e) : \tau}$$

(SKIP) $\gamma \vdash \text{skip} : H \text{ cmd}$

(ASSIGN) $$\frac{\gamma(x) = \tau \text{ var}, \quad \gamma \vdash e : \tau}{\gamma \vdash x := e : \tau \text{ cmd}}$$

(COMPOSE) $$\frac{\gamma \vdash c_1 : \tau \text{ cmd}, \quad \gamma \vdash c_2 : \tau \text{ cmd}}{\gamma \vdash c_1 ; c_2 : \tau \text{ cmd}}$$

(IF) $$\frac{\gamma \vdash e : \tau, \quad \gamma \vdash c_1 : \tau \text{ cmd}, \quad \gamma \vdash c_2 : \tau \text{ cmd}}{\gamma \vdash \text{if } e \text{ then } c_1 \text{ else } c_2 : \tau \text{ cmd}}$$

Figure 5 Typing rules.

(WHILE)
$$\frac{\gamma \vdash e : \tau, \quad \gamma \vdash c : \tau \text{ cmd}}{\gamma \vdash \text{while } e \text{ do } c : \tau \text{ cmd}}$$

(BASE) $L \leq H$

(REFLEX) $\rho \leq \rho$

(CMD$^-$)
$$\frac{\tau_1 \leq \tau_2}{\tau_2 \text{ cmd} \leq \tau_1 \text{ cmd}}$$

(SUBTYPE)
$$\frac{\gamma \vdash p : \rho_1, \quad \rho_1 \leq \rho_2}{\gamma \vdash p : \rho_2}$$

Figure 5 (Continued).

We begin with some definitions:

Definition 4.1: Memories μ and ν are equivalent with respect to a typing γ, written $\mu \sim_\gamma \nu$, if $\mu(h) = \nu(h)$ and $\mu(x) = \nu(x)$ for all x where $\gamma(x) = L$ var or $\gamma(x) = L$.

Definition 4.2: We say that c is a low command with respect to γ if the only occurrences of high variables in c with respect to γ are references to h in f(h).

Definition 4.3: Given a joint distribution d on dom(γ), we say that command c', is a low probabilistic simulation of a command c, relative to γ and d, if c' is a low command with respect to γ, and if c terminates in μ in k steps with unconditional probability q, relative to d, then there is a memory ν such that c' terminates in ν in at most $k+1$ steps with probability q', $q' \geq q$ and $\nu \sim_\gamma \mu$.

We need the following lemma:

Lemma 4.1: Suppose c is a well-typed command with respect to γ and that it has no occurrence of $f(h) = f(r)$. Then there is a low command c' with respect to γ such that for all μ where dom(μ) = dom(γ), whenever $(c, \mu) \overset{n}{\rightarrow} \mu'$, there is a μ'' and m such that $((c', \mu) \overset{m}{\rightarrow} \mu'', \mu' \sim_\gamma \mu''$ and $m \leq n$.

A proof of this lemma can be obtained by modifying the proof of Theorem 5.1 in Volpano and Smith (2000) in order to treat the slightly different notion of memory equivalence used here and to handle calls to f.

Finally, the reduction is given by the following theorem:

Theorem 4.2: *If c is a well-typed command with respect to* γ *and d is a joint distribution on dom(*γ*), then c has a low probabilistic simulation relative to* γ *and d.*

Proof. There are two cases, one where c has no instances of $f(h) = f(r)$ and the other where it does. First, suppose that c has no occurrence of $f(h) = f(r)$. Then let c' be the low command given by Lemma 4.1 for c. We can show that c' is a low probabilistic simulation of c as follows.

Let d be a joint distribution on dom(γ) and let

$$M = \{\mu \mid \mathrm{dom}(\mu) = \mathrm{dom}(\gamma)\}.$$

Suppose c terminates in a memory μ in k steps with unconditional probability q relative to d. Let $\mu_1, \ldots, \mu_n$ be all memories in M for which $(c, \mu_i) \overset{j}{\rightarrow} \mu$ for some j where $j \leq k$. Then

$$q = d(\mu_1) + d(\mu_2) + \square\square\square + d(\mu_n).$$

By Lemma 4.1, there is a μ'_i and m_i for each μ_i such that $(c', \mu_i) \overset{m_i}{\rightarrow} \mu'_i$, $\mu'_i \sim_\gamma \mu$ and $m_i \leq j$. Let ν_i be μ_i such that dom(ν_i) contains exactly h and all low variables of γ. Since c' is low, there is a ν'_i such that $(c', \nu_i) \overset{m_i}{\rightarrow} \nu'_i$, $\nu'_i \sim_\gamma \mu'_i$, and dom($\nu'_i$) = dom($\nu_i$), for $i = 1, \ldots, n$. By transitivity of $\sim_\gamma$,

$$\mu'_1 \sim_\gamma \mu'_2 \sim_\gamma \square\square\square \sim_\gamma \mu'_n.$$

Therefore, $\nu'_1 = \nu'_2 = \square\square\square = \nu'_n$.

So let $\nu = \nu'_1$. And c' terminates in ν in at most max $(m_1, \ldots, m_n)$ steps with unconditional probability at least q if for any memory ν', whose domain contains exactly h and all low variables of γ, c' begins execution in ν' with probability

$$\sum_{\{\mu \in M | \mu \sim_\gamma \nu'\}} d(\mu).$$

Also, $\max(m_1, \ldots, m_n) \leq k$.

Now suppose c has an occurrence of $f(h)=f(r)$. Without loss of generality, assume c has the form

$$\text{if } f(h) = f(r) \text{ then } c_1 \text{ else } c_2$$

where c_1 and c_2 have no instances of $f(h)=f(r)$. There is no loss of generality here because $f(h)=f(r)$ is constant in any memory, given that h and r are read-only variables in every typing, and c is well typed under γ. So let c'_1 and c'_2 be the low commands given by Lemma 4.1 for $c_1$and c_2, respectively. We can show that the command c' given by

$$(\text{if } X \text{ then } c'_1 \text{ else } c'_2); X := 0$$

where X is an independent Boolean random variable not in dom(γ), is a low probabilistic simulation of c.

Suppose, relative to d, that c_1 terminates in μ in fewer than k steps with probability q_1 given that $f(h)=f(r)$. Since there is no free occurrence of r in c_1, it also terminates in μ in fewer than k steps with probability q_1 given that $f(h)\neq f(r$. . Therefore, q_1 is an unconditional probability that c_1 terminates in μ in fewer than k steps. Likewise, suppose c_2 terminates in μ in fewer than k steps with probability q_2 given that $f(h)\neq f(r)$. Since there is no free occurrence of r in c_2, it also terminates in μ in fewer than k steps with probability q_2 given that $f(h)=f(r)$. So q_2 is an unconditional probability that c_2 terminates in μ in fewer than k steps. Therefore, c terminates in μ in k steps with probability

$$q = p\Box q_1 + (1-p)\Box q_2$$

where p is defined by

$$\sum_{\{\mu \in M | f(\mu(h)) = f(\mu(r))\}} d(\mu)$$

From above, there are memories ν'_1 and ν'_2, each equivalent to μ, such that c'_1 terminates in ν'_1 in fewer than k steps with probability q'_1, c'_2 terminates in ν'_2 in fewer than k steps with probability q'_2, $q'_1 \geq q_1$ and $q'_2 \geq q_2$. By the transitivity of $\sim_\gamma$, $\nu'_1 \sim_\gamma \nu'_2$ which implies $\nu'_1 = \nu'_2$ since neither has in its domain a high variable of γ besides h.

Now if $(c'_1, \nu_1) \xrightarrow{j} \nu'_1$, for some j and ν_1, then $(c'_1, \nu_1[X := n]) \xrightarrow{j} \nu'_1[X := n]$ because X does not occur free in c'_1. Likewise for c'_2. And if c' terminates, it does so in a memory that maps X to 0. Therefore, take $\nu = \nu'_1[X := 0]$, and we have that $\nu'_1[X := 0] \sim_\gamma \mu$ since $\nu'_1[X := 0] \sim_\gamma \nu'_1$ and $\nu'_1 \sim_\gamma \mu$. Finally, the unconditional probability that c' terminates in ν in at most $k+1$ steps is the probability that

$$\text{if } X \text{ then } c'_1 \text{ else } c'_2$$

terminates in ν'_1 in at most k steps. Because X is independent of dom(γ), this latter probability is given by

$$q' = p \Box q'_1 + (1-p) \Box q'_2$$

if for any memory ν', whose domain contains exactly h and all low variables of γ, command c' begins execution in $\nu'[X := 1]$ with probability

$$p \Box \sum_{\{\mu \in M | \mu \sim_\gamma \nu'\}} d(\mu)$$

Finally, $q' \geq q$. □

Suppose γ is a typing with a low variable l in its domain, d is a distribution on dom(γ) and c is a command for copying h to l that is well-typed relative to γ. Now suppose we run c simultaneously in all memories whose domains are equal to dom(γ) for $p(n)$ steps according to the input distribution d, where p is a polynomial and n is the length of the binary encoding of a memory. And suppose that after $p(n)$ steps, c terminates in a memory μ where $\mu(\mathrm{l}) = \mu(h)$ with probability q. By Theorem 4.2, there is a low command c' that terminates in no more than $p(n)+1$ steps in a memory ν where $\nu \sim_\gamma \mu$ with probability at least q. Furthermore, $\nu(\mathrm{l}) = \nu(h)$ since $\nu \sim_\gamma \mu$. And because c' is low, it has therefore managed to find h without any high variables as input, just occurrences of $f(h)$ is all. This brings us to the following Corollary:

Corollary 4.3: *Any bound on the probability of finding h from f(h) within polynomial time, for a particular integer size and distribution on h, also applies to the probability of leaking h with a well-typed command in polynomial time with respect to that size and distribution.*

Notice that probability p in the preceding proof takes into account the probability that $h = r$ as well as the collision resistance of f. Indeed, we would expect our simple password checker to be run with fairly high probability in a memory where $h = r$ if h stores a password and r is the checker's input. The reduction says that any well-typed program that attempts to exploit this fact has no advantage over a program that cannot reference h or r, but instead can access $f(h)$, call f on inputs of its choice and flip a coin. The one-wayness of f is treated by allowing instances of $f(h)$ in a low probabilistic simulation, which is a program squarely within the realm of a probabilistic model of computation used to define a one-way function (Sipser, 1997).

5. APPLICATION TO PASSWORD SYSTEMS

Consider again our simple password checker

$$b := (f(h) = f(r))$$

where variable h stores a password, b is an output variable and r is the input to the checker. Now we want to argue that the checker is secure. We begin by asserting what we know about the free variables. Well, since the output of the checker is public, we expect b to be low. On the other hand, h stores a password so it should be high. Under normal use of the checker, r will likely store the contents of h, and since h is high, we assert that r is high as well. Furthermore, the checker does not attempt to update h or r and therefore is well typed under the assumption that these variables are read-only. So the checker is secure in the sense that it belongs to a class of programs for which the complexity of leaking h rests upon the intractability of inverting $f(h)$ for sufficiently long and uniformly distributed values of h, by the above corollary. The checker's low probabilistic simulation is given by

$$(\text{if } X \text{ then } b := 1 \text{ else } b := 0);\ X := 0$$

where X is the random variable in the proof of Theorem 4.2.

Now suppose passwords are stored in a read-protected file in the clear as in, for example, a secrets file for CHAP (cryptographic handshake authentication protocol) widely used by PPP. In this case, the checker becomes just

$$b := (h = r)$$

We can argue that this checker too is secure using the reduction in Theorem 4.2 where we assume f is the identity function. But this assumption requires that rule IMAGE be eliminated, for clearly it is no longer sound. This means the adversary can no longer access the "resource" $f(h)$. Instead, we replace this form of access to h with a new form, namely match(h, e), which is true in memory μ if $\mu(h) = \mu(e)$. It has the following typing rule:

$$\frac{\gamma \vdash e : L}{\gamma \vdash \text{match}(h, e) : L}$$

Again, there is downgrading taking place, as in rule IMAGE. Whether match has any utility from the standpoint of writing useful programs is not important. What is important is that we provide the adversary with the resources we would realistically expect it to have. In the case of one-way functions, the adversary expects $f(h)$, but with f treated as the identity function, it now becomes the ability to match inputs of the adversary's choice against h which is precisely what match provides.

If access to h is limited to match queries and the values of h are uniformly distributed k-bit integers, then the probability of successfully leaking h with any deterministic polynomial-time command containing an independent random variable goes to zero as k increases (Volpano and Smith, 2000). If rule QUERY is replaced by the rule

$$\gamma \vdash h = r : L$$

```
if  f(h) = f(old) then
      check strength of new password
      h := new;
else  skip
```

Figure 6 A password update program for h.

then the second checker is well typed in the modified system, and is therefore secure in the sense that it belongs to a class of programs for which the complexity of leaking h rests upon this asymptotic hardness result, by Theorem 4.2.

Finally, to say something about the password system as a whole, we need to treat password updates as well. A password updater for h is given in Figure 6 .

The updater expects the old password, so free variable old is asserted to be a high variable, as is new which stores the new password. This program is well typed in the type system of Volpano et al. (1996), assuming the strength-checking portion is well typed. Therefore, it satisfies a noninterference property which is appropriate for this program, as there is no downgrading taking place.

The results here can also be applied to the GNU implementation of CHAP. The C code that hashes randomly generated server challenges with a shared CHAP secret, using RSA's MD5, is well typed. That tells us the code belongs to a class of programs for which leaking shared secrets is as hard as inverting 16-byte MD5 message digests computed over random challenges and sufficiently long and uniformly distributed CHAP secrets. It is really only in this sense that one can argue the code "protects" the confidentiality of shared CHAP secrets.

When modeling adversaries, we can identify essentially two kinds: inside and outside. Inside adversaries, write programs that we want to trust and have direct access to secrets like h and r. Outside adversaries, write programs we never trust, and therefore are denied direct access to secrets through some sort of access control. Each adversary has a typing rule where downgrading occurs. For the outside adversary, it is rule IMAGE (or the rule for match if f is the identity) and for the inside adversary, it is the rule for match or rule QUERY. Both kinds of adversary should be represented in a computational model.

6. CONCLUSION

This paper presents syntactic conditions, via a type system, for introducing one-way functions into a programming language with more flexibility than what noninterference allows. The conditions are sound in a computational sense and allow one to argue for the security of some systems where downgrading occurs.

Notice that functions are not part of the language we considered. That means commands in the language cannot call other commands. Functions pose a problem since λ-bound variables, although constant in a function body, can be bound in different ways through different function applications. This capability breaks the reduction. A useful line of work would be to identify conditions under which functions could be introduced securely.

ACKNOWLEDGMENTS

I would like to thank Geoffrey Smith for his comments on the paper. This material is based upon activities supported by the National Science Foundation under Agreement No. CCR-9900909.

REFERENCES

Kozen, D. (1981). Semantics of probabilistic programs. *Journal of Computer and System Sciences* 22:328–350.

Rivest, R. (1990). *Cryptography. Handbook of Theoretical Computer Science*. Chapter 13. Vol. A. The MIT Press/Elsevier.

Schneier, B. (1996) (1996). *Applied Cryptography*. 2nd. New York: John Wiley & Sons.

Sipser, M. (1997) (1997). *Introduction to the Theory of Computation*. PWS Publishing Company.

Volpano, D., Smith, G. (1999). Probabilistic noninterference in a concurrent language. *Journal of Computer Security* 7(2,3):231–253.

Volpano, D., Smith, G. (2000). Verifying secrets and relative secrecy. In: Proceedings 27th Symposium on Principles of Programming Languages. Boston, MA, Jan. pp. 268–276.

Volpano, D., Smith, G., Irvine, C. (1996). A sound type system for secure flow analysis. *Journal of Computer Security* 4(2,3):167–187.

15

Policy-Based Authentication and Authorization

Secure Access to the Network Infrastructure

JEFF HAYES
CISSP, Cereberian, Inc., Draper, UT, USA

1. OVERVIEW

A gaping hole in many of today's networks is the weak security surrounding the network devices themselves—the routers, the switches, and the access servers. In all public networks and in some private networks, the network devices are shared virtually among different user communities. Access to the configuration schemes and command lines is most often an "all or nothing" proposition—the network administrator gets either read-only privileges or read/write privileges. In this case, authentication equals authorization. Herein lies the problem.

Security policies may mandate certain administrators have read-only capabilities for all device parameters and read/write capabilities for a certain subset of commands. Each administrator may have a unique access profile. Authentication verifies identity. Authorization verifies privileges. This paper will address the value of using a centralized provisioned management structure that disseminates network policies and administration privileges to all the devices that make up the network infrastructure.

2. AUTHENTICATION

With the mission critical nature of today's LANs, businesses cannot afford to have their data networks compromised by unauthorized users. Until now, the main security device or implementation for network access has been the firewall or router-based access control lists. Providing a barrier between untrusted networks (like the Internet) and internal, trusted networks is important but it does not end there.

Security experts today warn that while the external threat to networks is real, the largest threat often comes from inside the company. Authentication of internal users has long been established as the primary security device for file servers, network operating systems, and mainframes. There are also authentication requirements for routing tables (RIP, OSPF, BGP4), switch ports, router and switch configuration files, and web servers, to name but a few.

Traditional authentication—userID and password submitted in clear text—is typically not adequate for most security policies. People tend to use simple passwords or write them down in view of a potential perpetrator. Passwords can be stolen, sniffed, guessed, attacked via freeware dictionary tools or brute force attacks, compromised though insecure password files, and obtained through social engineering. Some passwords never expire. Some expire every 60–90 days without allowing the user to reuse an old password. Some are short, simple alpha characters only. Others are a combination of alpha, numeric, and special characters. Some password files are stored in clear text; others are encrypted. Many are transmitted in clear text; others as cipher text. Whatever the method, some identification is better than none.

An area often overlooked is the authentication associated with the network infrastructure—the routers, switches, and access servers. The same issues associated with network operating systems and user applications have bearing in the infrastructure. Some methods are strong, while others are weak. The idea presented here is to distribute device and user authentication to these devices to a standalone authentication server, as opposed to storing the information on each device independently (see Figure 1).

For the past decade, the IT industry has seen an evolution in authentication techniques. Though most users rely on a user ID and password to establish their identity, more reliable authentication schemes involve multiple factors,

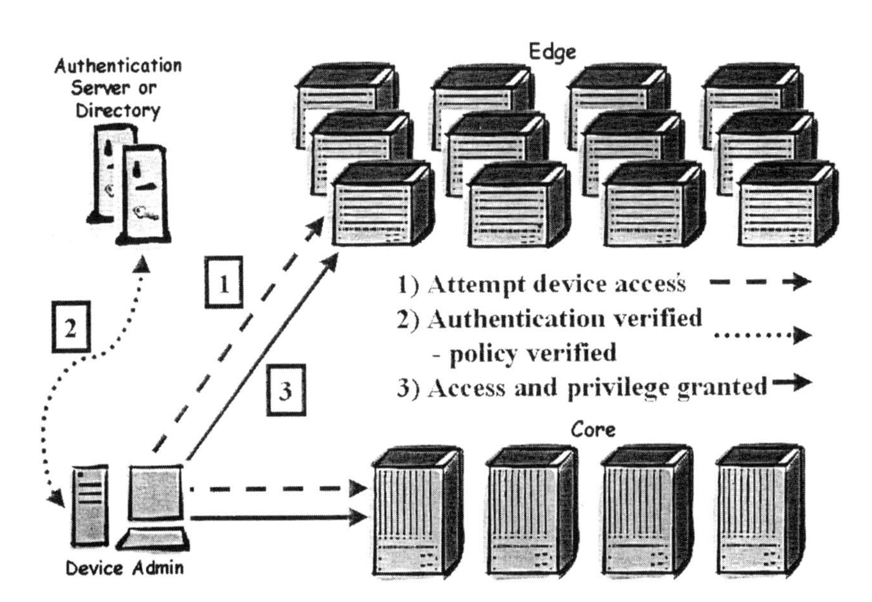

Figure 1 Authenticated device access.

insuring a great chance of accurate identification. These factors include:

- Something you are, a biometric characteristic that is unique to every human. Fingerprints, hand prints, faces, retinas, voices, and keystroke timing can all be tied to a unique individual.
- Something you know, user ID and password. It is currently the most widely used form of identification.
- Something you possess, which typically involves external security devices including banking/ATM cards, tokens, and smart cards.

Advanced multiple-factor authentication techniques are needed to provide assurance that the user desiring connectivity is who she claims to be. There are a number of key methods for implementing this level of authentication.

One-time password schemes provide authentication over unsecured networks. The schemes can be based on one of two systems: (1) passwords stored both on a client device and on a central server; or (2) passwords kept on a central system and requested on demand by users. Because each password is only used once, most are sent in clear, unencrypted text, for example SASL methods like S/KEY.

Time-based passwords are based on both a password and an external security device. Users desiring access possess a hand-held device or token. When prompted to log in they identify themselves with an ID and a one-time password that is displayed on the token. The resulting password is a combination of a PIN and the number generated by the device's LCD. The users' temporary LCD number is synchronized with a central authentication server. If they match, the user is authenticated. An example is an RSA' SecurID token and ACE/Server.

Challenge and response systems are also two-factor authentication systems that leverage hand-held devices. The initial login request causes the authentication, server to generate a random numeric challenge. Users unlock their hand-held device by punching in a PIN on the card's keypad. Users enter the challenge into the card as it was received.

The card uses a known algorithm, like Data Encryption Standard (DES), to calculate and display the response to the challenge. Users enter the card's response to complete the login process. CRAM is the common technology used for this. Smart cards are similar to the aforementioned token systems but contain more intelligence and processing power—small microprocessors with embedded encryption. Smart cards communicate directly with the authentication server through a card reader. Users provide the initial PIN and the card does the rest—exchanges keys and agrees on the encryption algorithms to be used.

These authentication technologies are typically complemented by services and/or servers that facilitate user profile management. The following authentication services add the element of authorization to the authorization process, something not provided by the above solutions. Remote Access Dial In User Service (RADIUS) systems use a client or agent to facilitate users' login requests. The client passes the information to a RADIUS server, which authenticates the user. All communication between the client and server is authenticated and not sent in clear text. Servers can support a variety of authentication methods including PAP, CHAP, UNIX login, time-based tokens, and challenge/response systems. RADIUS is widely used by ISPs and other dial-up applications. RADIUS is also being used for user authentication, authorization, and accounting beyond dial-up applications, including LAN-based applications. As such, a new standard known as DIAMETER is being proposed that attempts to expand on RADIUS' known shortcomings, resulting in a broader protocol.

X. 500 Directory Servers with X. 509 using either simple (passwords) or strong (public-key) authentication accessible via the Lightweight Directory Access Protocol (LDAP) are becoming a critical information repository for end user profiles. Most of X. 509 security elements are provided by RSA's Public Key Cryptography Standards (PKCS), although other methods exist. As network administrators see the value of minimizing the number of directories, there will be a move to consolidate directories and/or to utilize the meta-directory concept.

The Burton Group defines a meta-directory service as being a class of enterprise directory tools that integrate existing, or "disconnected", directories by addressing both the technical and political problems inherent in any large-scale directory integration project. A big challenge albeit a worthy one.

Kerberos is a strong, single sign-on authentication system with a goal of validating a principal's identity. A principal can be a client, a program, or a service. For each service, a client needs, a server generates a ticket for that client/server server session. Each ticket contains a number of components: client, server, address, time stamp, lifetime, key (c/s), and key (s). Kerberos is a published standard and is a true single sign-on technology—user logs in once and gains access to all pre-authorized resources without requiring a new or re-entered password. Kerberos is in use in many environments, namely North American colleges and universities. With Windows 2000 using Kerberos v5 as its default network authentication protocol, Kerberos may now become mainstream, albeit Microsoft's version of mainstream.

For many, especially in the enterprise, the idea of single sign-on is a network panacea. But given efforts by standards groups (GSS-API and CDSA) and individual companies like Novell (NICI), Microsoft (SSPI and CryptoAPI) and Sun (Java 2), it appears it will be some time before homogenous authentication will be a reality.

3. AUTHORIZATION

Authorization is the granting of privileges associated with an authenticated user, device, or host. The traditional way authorization is granted is exemplified by common operating systems like UNIX. A super-user is the all-powerful system owner or administrator. That individual has the authority to grant privileges to other users. These privileges can be "read", "write", "append", "lock", "execute", and "save"—individually or in combination. Less traditional although analogous are the privileges granted to network administrators—those that manage the network infrastructure.

A network is comprised of many different devices—from host machines to application, file, web, DNS, and communication servers; from remote access servers to hubs and workgroup switches on the edge; and from WAN-oriented routers to LAN- or ATM-oriented core switches and routers. There is a growing need to grant privileges to these systems on a need-to-know basis.

In order to permit this, the network devices must be able to support a provisioned management structure. The privileges that can be granted could be broken down into devices, services, and configuration parameters.

Device access security is analogous to tradition access control list or firewall rules. The security administrator creates specific rules that limit access to the network devices based on characteristics of the device requesting access, for example, source and/or destination IP address or source MAC address. This traditional access control concept keeps all of the authorization on the device itself. The provisioned management structure proposed here ties on-device authentication and authorization rules to an external directory server. Access is granted provided the policy allows for it.

For example, an IP source (host or network) attempts to access an IP destination (host or network). The network device recognizes a policy exists for this request—it has a matching rule. It queries the directory to determine what to do with it. The appropriate policy is returned to the device and implemented accordingly. Besides this implicit application, the policy could also be associated with explicit information like time-of-day or -month (Figure 2).

Services-based access involves management protocols like telnet, FTP, TFTP, HTTP, and SNMP. Much like what is listed above for device security, it may be prudent to allow certain users access only to specific services based on pre-defined policies. An example of how privileges can be allocated within a group of administrators is shown in Figure 3.

Configuration parameters are the tasks administrators are allowed to perform once they have been authenticated and granted access to the device. Similar to what is described above for service privileges, policy may exist that only allocates

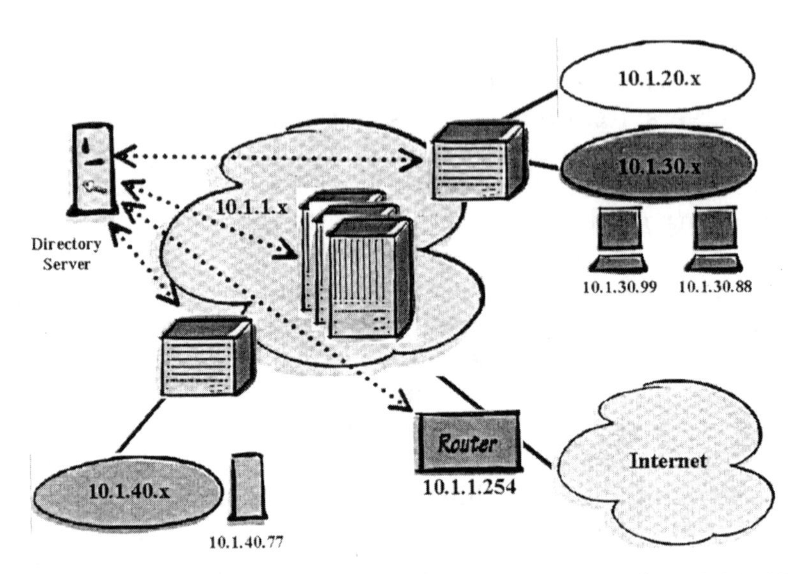

Figure 2 Access Rules: 10.1.20.x to 10.1.40.x = okay 8:00–18:00; Internet to 10.1.40.77 = okay after authentication; 10.1.30.99 & 10.1.30.88 to Internet = never.

management privileges to certain individuals based on job descriptions. For example, policy may dictate that only the super user and the two security managers have the ability to add, remove, or change user security profiles or privileges (Figure 4).

Name	User	HTTP	Telnet	FTP	TFTP	SMTP	Console	Custom
Robert Ivins	X	X	X	X	X	X	X	X
Julio Lopez	X	X	X	X	X	X	X	
Alain Chadoin	X	X	X	X	X		X	
Steve Allison	X					X		
Barb Wheeler	X	X	X	X	X			
Marie Roth	X					X		
Jay Ng	X	X	X	X	X		X	

Figure 3 Service privileges.

Name	Super User	Security	Routing	VLAN	WAN	QOS	ATM	Custom
Robert Ivins	X	X	X	X	X	X	X	X
Julio Lopez		X						X
Alain Chadoin					X	X	X	
Steve Allison			X	X				
Barb Wheeler		X	X	X	X	X	X	
Marie Roth					X		X	
Jay Ng			X	X				

Figure 4 Configuration privileges.

These configuration submenus or individual command privileges are allocated and stored in a common directory. These access accounts can contain both implicit and explicit rules ranging from device identifiers, network IDs, TCP or UDP ports, as well a configuration submenu command (CLI) privileges.

There are significant reasons for this fine-grained provisioning profile model. Most network devices have a combination of read-only and read/write privileges. In many cases, especially on tightly controlled enterprise networks, this is adequate. But as networks become more complex and as autonomous systems begin to share the same devices, there is a need to segment the administrative privileges into groups. This is magnified when network provisioning and management is out-sourced.

Many organizations look to external resources for assistance at managing their network edges and access. Managed services are a multi-billion dollar (US) business. Managed service providers are offering high-speed access to corporate Intranets, connecting common business partners via Extranet designs, or providing direct connection to the Internet to multiple tenants from a single device.

In the case of multi-tenant network access, the service provider may want to give each customer some basic trouble-shooting capabilities, but for their subnetwork only. One tenant should not be able to see anything relative to anyone else's network. In fact, they should have no idea that other tenants are sharing the same local access device. In addition, the service provider may not want to give its own employees free access to the device. For example, it may be proper to give most of the operations team read-only, routing, and VLAN configuration privileges. Other policy may only give the privilege of changing QOS parameters to a few individuals (Figure 5).

In order for authorization policies to be disseminated to the network devices, the network must be able to support a central repository for these policies. The network devices must be able to access those policies based on some event or pre-provisioned rule. The network can then decide what to do with the event in question. The device must also have the ability to enforce the policy. This policy deployment scenario is referred to as policy-based network management.

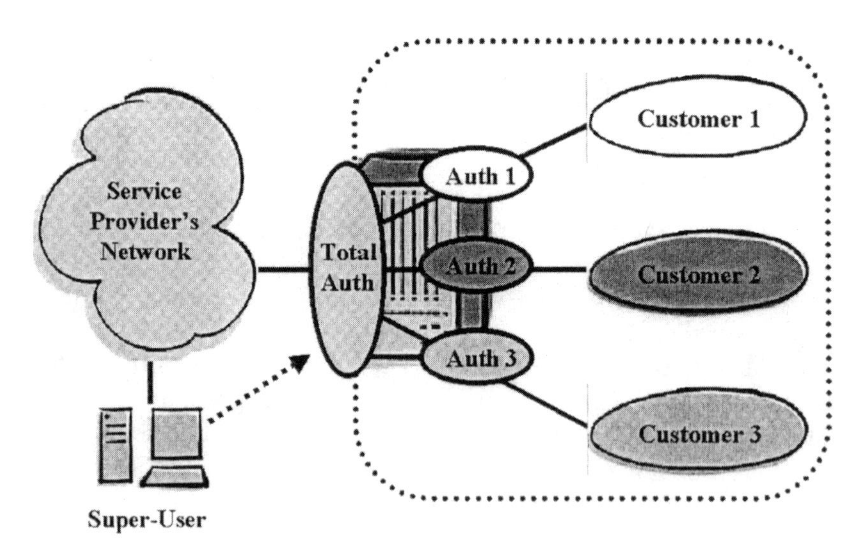

Figure 5 Single building–Multiple tenants.

4. POLICY MANAGEMENT

The power of this provisioned management structure is magnified when authentication and authorization are coupled with a centralized directory or policy server. Conceptually, when an administrator authenticates to the network, he/she is granted the ability to access all of the devices, services, and configuration parameters he/she has been pre-authorized to access. Each time the administrator attempts to access a network device, that device will query the policy server. The policy server will send an acknowledgement to the device granting authorizations for the requested service.

Policy-based network management leverages directories, the central repositories for policies. This is done for a very good reason. Instead of configuring each device with specific privileges, the devices consult the central directory for this information. This simplifies administration—instead of changing authentication and authorization information on dozens or hundreds of devices, it is done at a central location.

Policy-based management implementations, that leverage directory and policy servers, are offered by many vendors including Alcatel, Cisco, Avaya, Extreme, Foundry, and Nortel. All share a common design. They are all based on the concept of a policy console, a policy server or repository, a policy decision point (PDP), and a policy enforcement point (PEP) (Figure 6).

Policies can be recalled via some triggered event or it can be provisioned. In the case of the former, an event can be the arrival of a frame that the network device is unsure how to treat. For example, an IP source or destination address, MAC address, or IP multi-cast membership record can be the trigger. If the network device has no cached policy for that event, it must query the PDP. The PDP receives its policies from the directory server which are configured and stored there by a policy administrator via policy console downloads. The PDP informs the network device—which is the PEP—to follow specific policy instructions. The PEP implements the policy for that frame and related session flow.

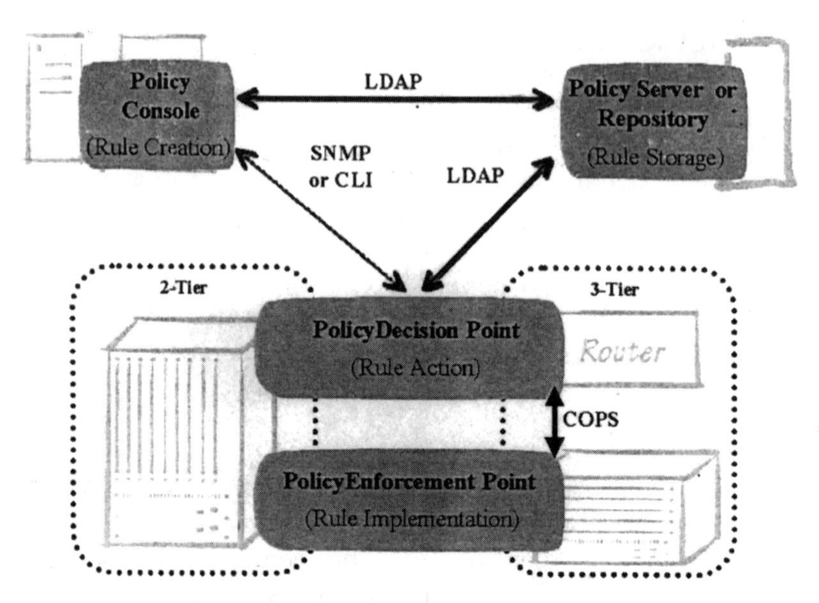

Figure 6 Policy–Management architecture

These policy management architectures are either a two-tiered or three-tiered design. A two-tiered method combines the PDP and PEP in the same network device. The three-tiered method has the PDP and PEP running in separate devices. The protocols used to communicate policies will depend on the newness of the products. For example, in newer gear, a separate PDP communicates to the PEP via the Common Open Policy Service (COPS) protocol. In an integrated PDP/PEP, the policy is communicated from the policy repository via LDAP. In older networking gear, the policy communication may be SNMP or CLI.

When the questions of availability and scalability are asked, the provisioned device management structure provides a positive response. Depending on the value placed on the network and its availability, repositories and servers can be redundantly implemented. In addition, based on the number of devices that will be accessing policies and the volume of policy decisions that will need to be made, scalability can be designed into the implementation.

An example of how a policy-based management implementation works is presented below. In this scenario, the triggered event (3) is an IP source address. The router has a rule (ACL) that states it must check with the policy server (4) in order to know how (or if) it should be forwarded. The PDP compares the request with the policy (obtained previously in (1) and (2)). Once it knows this information, it informs the router how the policy should be enforced (5). The traffic is then forwarded based on the policy (6).

The most effective manner this provisioned management structure can operate in is when the policy server, PDP and/or PEP, understand the concept of "state". State is best described as an awareness of network communications and the rules that are regulating it. State tables contain information like who logged on, when, and which resources are being accessed. Few policy or directory protocols understand the concept of state (COPS however does). For wide-scale usage, maintaining authentication and authorization state is a pre-requisite (Figure 7).

5. POLICY SECURITY

Communications between the policy console, policy repository, PDP and PEP must address security. It is becoming unacceptable practice to communicate device configuration profiles and parameters across the network in clear text. Adequate technology exists to allow this communication to be secured.

Secure Socket Layer (SSL) and its cousin Transport Layer Security (TLS) are widely used transport protocols that, when couple with public-key cryptography, provide a secure communications tunnel between clients and servers or between network devices. However, there is no assurance the user behind the client computer is an authenticated user.

Simple Authentication and Security Layer (SASL) is a standard-based, general authentication mechanism which can be used by any connection-oriented protocol, like SNMP, LDAP, and S/KEY. Digest Authentication is also an SASL mechanism used to secure HTTP communications, albeit less secure than others like SSL.

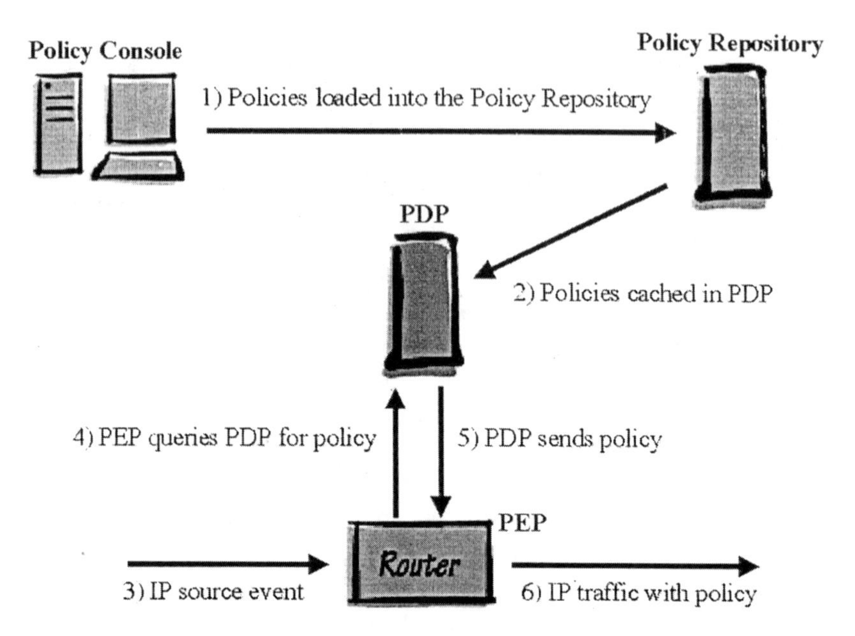

Figure 7

The best method for secure communications to/from/ between the devices that make up the network infrastructure will be a fully implemented Public Key Infrastructure (PKI) based on X. 509 authentication foundations and a standard-based family of encryption capabilities like RSA's PKCS. The issue with this model is it relies too heavily on a single vendor, RSA. However, because the RSA protocols are platform independent and considered technically sound, their appeal is wide. There is plenty of activity by other vendors to attempt to standardize PKI, without forcing vendors and the end users to pay RSA fees.

6. CONCLUSION

How much security is enough? How much is not enough? This proposal about how one can use a provisioned management structure for the network infrastructure is only useful to the organization that understands the value of its network and the information contained therein. For many, this management model is overkill. For others, it is well suited. Whatever the desire, organizations must understand the value of their networks and calculate the cost to the business if the network were unavailable.

The result of this assessment should be a corporate security policy document. This document will be the plan that a company will follow for all its security issues. It will clearly spell out the business values (strengths) and weaknesses (vulnerabilities). It will delineate what is important and what is not. From this, the corporate security budget, procedures, technologies, actions, and awareness programs can be deployed. Hopefully, requiring secure access to the network infrastructure will be part of the corporate information security agenda.

16

The Evolution of Viruses and Worms

THOMAS M. CHEN
Department of Electrical Engineering,
Southern Methodist University,
Dallas, TX, USA

JEAN-MARC ROBERT
Alcatel Canada Inc., Ottawa,
Ontario, Canada

ABSTRACT

Computer viruses and network worms have evolved through a continuous series of innovations, leading to the recent wave of fast-spreading and dangerous worms. A review of their historical development and recent outbreaks leads to a number of observations. First, while viruses were more common than worms initially, worms have become the predominant threat in recent years, coinciding with the growth of computer networking. Second, despite widespread use of firewalls and other network security equipment, worm outbreaks still occur and will likely continue to be a threat for the near future. Third, recent worms are appearing as a series of quick successive variants. Unlike the independent efforts of early viruses, these variants suggest an increasing level of coordination among worm creators. Fourth, recent worms have shown capabilities to spread faster and exploit more infection vectors. This trend implies a more urgent need for automated, coordinated protection measures. Finally, more dangerous payloads are becoming commonplace. This suggests that worm creators are using worms for other objectives than simply infection, such as data theft and setting up denial of service networks.

1. INTRODUCTION

Computer viruses and worms are characterized by their ability to self-replicate. The modern computer virus was conceived and formalized by Fred Cohen as a USC graduate student in 1983. Cohen wrote and demonstrated the first documented virus in November 1983 (Cohen, 1987). Like biological viruses, computer viruses reproduce by taking advantage of the existing environment. A biological virus consists of single or double-stranded nucleic acid (DNA or RNA) surrounded by a protein shell (capsid). The capsid gives specificity to bond with those particular hosts with matching surface receptors, while the inner nucleic acid gives infectivity or potency to subvert the infected host's cellular machinery. A virus is incomplete and inactive outside of a living cell but becomes active within a host cell by taking over the host's metabolic machinery to create new virus particles that spread the infection to other cells.

Computer viruses replicate themselves by attaching their program instructions to an ordinary "host" program or document, such that the virus instructions are executed during the execution of the host program. As a simplification of Cohen's definition, a basic computer virus can be viewed as a set of instructions containing at least two subroutines attached somehow to or within a host program or file (Cohen, 1994). The first subroutine of the virus carries out the infection by seeking out other programs and attaching or overwriting a copy of the virus instructions to those programs or files (Grimes, 2001). The method of infection or propagation is called the "infection vector" (borrowed from epidemiology). The second subroutine of the virus carries the "payload" that dictate the actions to be executed on the infected host. The payload could be almost anything in theory, for example, deletion of data, installation of backdoors or DoS (denial of service) agents, or attacks on antivirus software (Ludwig, 1998). An optional third subroutine could be a "trigger" that decides when to deliver the payload (Harley et al., 2001).

Computer networks have created a fertile environment for worms, which are related to viruses in their ability to self-replicate but are not attached parasitically to other programs. Worms are stand-alone automated programs designed to exploit the network to seek out vulnerable computers to infect with a copy of themselves. In contrast to viruses, worms are inherently dependent on a network and not dependent on any human action (such as to execute a program infected with a virus). Worms have become more prevalent since Internet connectivity has become ubiquitous. The Internet increases the vulnerability of all interconnected computers by making it easier for malicious programs computers to move among computers.

Recent worm outbreaks, such as the Blaster worm in August 2003 and the SQL Sapphire/Slammer worm in January 2003, have demonstrated that networked computers continue to be vulnerable to new attacks despite the widespread deployment of antivirus software, firewalls, intrusion detection systems, and other network security equipment. We review the historical development of viruses and worms to show how they have evolved in sophistication over the years. The history of virus and worm evolution is classified into four "waves" spanning from the Shoch and Hupp worms in 1979 to the present day (the term "generations" is less preferred because viruses and worms of one wave are not direct descendents from an earlier wave). This historical perspective leads to a number of observations about the current state of vulnerability and trends for possible future worm attacks.

We classify the evolution of viruses and worms into the following waves:

- first wave from 1979 to early 1990s;
- second wave from early 1990s to 1998;
- third wave from 1999 to 2001;
- fourth wave from 2001 to today.

These waves represent periods where new technological trends began and appeared in a significant number of cases. For example, the third wave was dominated by so-called mass e-mailers: viruses and worms that exploited e-mail programs to spread. The classification dates are approximate and not meant to imply that waves can be separated so distinctly. For example, mass e-mailers characteristic of the third wave are still common during the fourth wave.

2. FIRST WAVE: EARLY VIRUSES AND WORMS

The first wave of viruses and worms from roughly 1979 to early 1990s were clearly experimental. The early viruses were commonly boot-sector viruses and targeted mostly to MS DOS. The early worms were prone to programming bugs and typically hard to control. The term "worm" was created by John Shoch and Jon Hupp at Xerox PARC in 1979, inspired by the network-based multisegmented "tapeworm" monster in John Brunner's novel, *The Shockwave Rider*. Shoch and Hupp used worm to refer to any multisegmented computation spread over multiple computers. They were inspired by an earlier self-replicating program, Creeper written by Bob Thomas at BBN in 1971, which propelled itself between nodes of the ARPANET. However, the idea of self-replicating programs can be traced back as early as 1949 when the mathematician John von Neumann envisioned specialized computers or "self-replicating automata" that could build copies of themselves and pass on their programming to their progeny (Von Neumann and Burles, 1966).

Shoch and Hupp invented worms to traverse their internal Ethernet LAN seeking idle processors (after normal working hours) for distributed computing (Shoch and Hupp, 1982). Since the worms were intended for beneficial uses among cooperative users, there was no attempt at stealth, intrusion, or malicious payload. Even under cooperative conditions, however, they observed that worm management was a key problem to ensure that worm growth could be reliably contained. Their worms were designed with limited lifetimes, and responsive to a special packet to kill all worms. Despite these safeguards, one of the worm programs mysteriously ran out of control and crashed several computers overnight.

In 1983, Fred Cohen conceived, wrote and demonstrated the first computer virus while a graduate student at USC. The 1986 DOS-based Brain virus, supposedly written by two Pakistani programmers, was interesting in its attempt to stealthily hide its presence by simulating all of the DOS system calls that normally detect viruses, causing them to return information that gave the appearance that the virus was not there.

In 1987, the "Christma Exec" virus was among the first to spread by e-mail among IBM mainframes. It is also an early example of social engineering where the user is tricked into executing the virus because it promised to draw a Christmas tree graphic. The worm does produce a Christmas card graphic on the computer screen (drawn using a scripting language called Rexx) but also sends a copy of itself in the user's name to his list of outgoing mail recipients. The recipients believe the e-mail is from the user so they open the e-mail.

On November 2, 1988, the famous Morris worm disabled 6000 computers in a few hours (constituting 10% of the Internet at that time) (Spafford, 1989). The worm was written by a Cornell student, Robert Morris Jr. Investigations conducted later (ending in his conviction in 1990) concluded that his motives for writing the worm were unknown but the worm was not programmed deliberately for destruction. Instead, the damage appeared to be caused by a combination of accident and programming bugs. It was detected and contained only because an apparent programming error caused it to re-infect computers that were already infected, resulting in a noticeable slowdown in infected computers. It was among the first to use a combination of attacks to spread quickly: cracking password files; exploiting the debug option in the Unix "sendmail" program; and carrying out a buffer overflow attack through a vulnerability in the Unix "finger" daemon program.

In October 1989, the WANK (Worms Against Nuclear Killers) worm apparently learned from the Morris worm and infected VMS computers on DECnet. It spread using e-mail functions, exploited default system and field service accounts and passwords for access, and tried to find accounts where the user name and password were the same or the password was null.

3. SECOND WAVE: POLYMORPHISM AND TOOLKITS

The second wave in the approximate period from the early 1990s–1998 saw much more activity in viruses than worms, although the technical advances in viruses would effect the later evolution of worms. In this period, viruses began to move from Microsoft DOS to Windows as the primary target; cross-platform macroviruses appeared; mass creation of polymorphic viruses became easy; and a trend towards e-mail as the preferred infection vector began.

In the late 1980s, the idea of using encryption to scramble the appearance of a virus was motivated by the fact that antivirus software could detect viruses by scanning files for unique virus signatures (byte patterns). However, to be executable, an encrypted virus must be prepended with a decryption routine and encryption key. The decryption routine remains unchanged and therefore detectable, although the key can change which scrambles the virus body differently. Polymorphism carries the idea further to continuously permute the virus body. A polymorphic virus reportedly appeared in Europe in 1989. This virus replicated by inserting a pseudorandom number of extra bytes into a decryption algorithm that in turn decrypts the virus body. As a result, there is no common sequence of more than a few bytes between two successive infections.

Polymorphism became a practical problem in 1992 when a well-known hacker, Dark Avenger, developed a user-friendly Mutation Engine to provide any virus with polymorphism (Smith, 1994). Other hackers soon followed with their own versions of mutation engines with names such as TPE, NED, and DAME. In 1994, Pathogen and Queeg were notable polymorphic DOS-infecting viruses that were produced by Black Baron's SMEG (Simulated Metamorphic Encryption engine).

The "virus creation lab" was a user-friendly, menu-driven programming toolkit that allowed hackers with little programming skill to easily generate hundreds of new viruses. Other virus creation toolkits soon followed such as PS-MPC. Perhaps, the best known product of a virus toolkit was the 2001 Anna Kournikova virus. This virus was carried in an e-mail attachment pretending to be a JPG picture of the tennis player. If the Visual Basic Script attachment is executed, the virus e-mails a copy of itself to all addresses in the Outlook address book.

The 1995 Concept virus was the first, written for Word for Windows 95. It infected Word's "normal.dot" template so that files were saved as templates and ran the infective AutoOpen macro. An infected file was accidentally shipped on a Microsoft CD called "Microsoft Windows95 Software Compatibility Test". Later, Microsoft UK shipped an infected file on another CD called "The Microsoft Office 95 and Windows95 Business Guide". The vast majority of macro-viruses are targeted to Microsoft Office documents. Macro-viruses have the advantages of being easy to write and cross-platform. However, most people now know to disable macros in Office so macroviruses have lost their popularity.

4. THIRD WAVE: MASS E-MAILERS

The third wave spanning roughly 1999 to late 2000 is highlighted by a large number of mass e-mailers, beginning with the "Happy99/Ska" worm. E-mail continues to be a very popular infection vector today. In January 1999, the Happy99 worm spread by e-mail with an attachment called Happy99.exe. When the attachment is executed, it displayed fireworks on the screen to commemorate New Year's Day 1999, but secretly modified the WSOCK32.DLL file (the main Windows file for Internet communications) with a Trojan horse program that allowed the worm to insert itself into the Internet communications process. The original WSOCK32.DLL file is renamed to WSOCK32.SKA. Every e-mail sent by the user generated a text-less second copy that carried the worm to the same recipients.

In March 1999, the Melissa macrovirus spread quickly to 100,000 hosts around the world in 3 days, setting a new record and shutting down e-mail for many companies using Microsoft Exchange Server. It began as a posting on the Usenet newsgroup "alt.sex" promising account names and passwords for erotic web sites. The attached Word document actually contained a macro that used the functions of Microsoft Word and the Microsoft Outlook e-mail program to propagate. Up to that time, it was widely believed that a computer could not become infected with a virus just by opening e-mail. When the macro is executed in Word, it first checks whether the installed version of Word is infectable. If it is, it reduces the security setting on Word to prevent it from displaying any warnings about macrocontent. Next, the virus looks for a certain Registry key containing the word "kwyjibo" (apparently form an episode of the television show, "The Simpsons"). In the absence of this key, the virus launches Outlook and sends itself to 50 recipients found in the address book. Additionally, it infects the Word "normal.dot" template using the VBA macro auto-execute feature. Any word document saved from the template would carry the virus.

The PrettyPark worm became widespread in the summer of 1999. It propagates as an e-mail attachment called "Pretty Park.exe." The attachment is not explained but bears the icon of a character from the television show, "South Park." If executed, it installs itself into the Windows system directory and modifies the registry to ensure that it runs whenever any

.EXE program is executed. This can cause problems for antivirus software that runs as an .EXE file. Furthermore, the worm e-mails itself to addresses found in the Windows Address Book. It also transmits some private system data and passwords to certain IRC (Internet relay chat) servers. Reportedly, the worm also installs a backdoor to allow a remote machine to create and remove directories, and sent, receive, and execute files.

In June 1999, the ExploreZip worm appeared to be a WinZip file attached to e-mail but was not really a zipped file. If executed, it would display an error message, but the worm secretly copied itself into the Windows systems directory or loaded itself into the registry. It sends itself via e-mail using Microsoft Outlook or Exchange to recipients found in unread messages in the inbox. It monitors all incoming messages and replies to the sender with a copy of itself.

In early 2000, the BubbleBoy virus (apparently named from an episode of the television show, "Seinfeld") demonstrated that a computer could be infected just from previewing e-mail without necessarily opening the message. It took advantage of a security hole in Internet Explorer that automatically executed Visual Basic Script embedded within the body of an e-mail message. The virus would arrive as e-mail with the subject "BubbleBoy is back" and the message would contain an embedded HTML file carrying the viral VB Script. If read with Outlook, the script would be run even if the message is just previewed. A file is added into the Windows startup directory, so when the computer starts up again, the virus e-mails a copy of itself to every address in the Outlook address books. Around the same time, the KAK worm spreads by a similar exploit.

In May 2000, the fast-spreading Love Letter worm demonstrated a social engineering attack, which would become common in future mass e-mailing worms (CERT advisory, CA-2000-04). It propagates as an e-mail message with the subject, "I love you" and text that encourages the recipient to read the attachment. The attachment is a Visual Basic Script that could be executed with Windows Script Host (a part of Windows98, Windows2000, Internet Explorer 5, or

Outlook 5). Upon execution, the worm installs copies of itself into the system directory and modifies the registry to ensure that the files were run when the computer started up. The worm also infected various types of files (e.g., VBS, JPG, MP3, etc.) on local drives and networked shared directories. When another machine is infected, if Outlook is installed, the worm will e-mail copies of itself to anyone found in the address book. In addition, the worm makes an IRC connection and sends a copy of itself to anyone who joins the IRC channel. The worm had a password-stealing feature that changed the startup URL in Internet Explorer to a web site in Asia. The web site attempted to download a Trojan horse designed to collect and e-mail various passwords from the computer to an address in Asia.

In October 2000, the Hybris worm propagated as an e-mail attachment (CERT incident note, In-2001-02). If executed, it modifies the WSOCK32.DLL file in order to track all Internet traffic. For every e-mail sent, it subsequently sends a copy of itself to the same recipient. It is interesting for its capability to download encrypted plug-ins (code updates) dynamically from the “alt.comp.virus” newsgroup. The method is sophisticated and potentially very dangerous, since the worm payload (destructive capability) can be modified at any time.

5. FOURTH WAVE: MODERN WORMS

The fourth wave of modern worms began in 2001 and continues today. They are represented by worms such as Code Red and Nimda that demonstrate faster spreading and a new level of sophistication including

- blended attacks (combined infection vectors);
- attempts at new infection vectors (Linux, peer-to-peer networks, instant messaging, etc.);
- dynamic code updating from the Internet;
- dangerous payloads;
- active attacks against antivirus software.

Linux was first targeted by the Bliss virus in 1997. In early 2001, Linux was hit by the Ramen and Lion worms.

The Lion worm seemed to be unusually dangerous. After infection of a new victim, the worm carries out several actions. The worm sends e-mail containing password files and other sensitive information, e.g., the IP address and username of the system and bundles the contents of the "/etc/password," "/etc/shadow," and "/sbin/ifconfig" files into the mail.log file before transmission. The worm installs the binary toolkits "t0rn" and the distributed denial of service (DDoS) agent "TFN2K" (Tribal Flood Network 2000). The t0rn rootkit deliberately makes the actions of the worm harder to detect through a number of system modifications to deceive the utility "syslogd" from properly capturing system events. The worm creates a directory named "/dev/.lib" and copies itself there. The "etc/host.deny" file is deleted to make any protection offered by TCP Wrappers useless against an outside attack. The worm also installs backdoor root shells on TCP port 60008 and TCP port 33567. A Trojanized version of SSH on port 33568 changes the port setting to listening. This port setup could be done to launch future DDoS attacks using the TFN2K agent.

In February 2001, the Gnutelman/Mandragore worm infected users of Gnutella peer-to-peer networks by disguising itself as a searched file.

In May 2001, the Sadmind worm spread by targeting two separate vulnerabilities on two different operating systems, and set a precedent for subsequent viruses that could combine multiple attacks (CERT advisory, CA-2001-11). It first exploited a buffer overflow vulnerability in Sun Solaris systems and installed software to carry out an attack to compromise Microsoft IIS web servers.

A buffer overflow vulnerability in Microsoft's IIS web servers was announced on June 18, 2001, referred to as the Index Server ISAPI vulnerability (CERT incident note, IN-2001-08). About a month later on July 12, 2001, the first Code Red I (or .ida Code Red) worm was observed to exploit this vulnerability. Upon infection, the worm set up itself in memory and generated up to 100 new threads, each an exact replica of the original worm. The first 99 threads were assigned to spread the worm by infecting other IIS servers.

The worm generated a pseudorandom list of IP addresses to probe but used a static seed which (unintentionally) resulted in identical lists of IP addresses generated on each infected host. As a result, the spread was slow because probes repeatedly hit previously probed or infected machines, although 200,000 hosts were infected in 6 days. A side effect was congestion between these hosts, causing a DoS effect. The last 100th thread was assigned to check the language of the IIS system; if English, then the worm would deface the system's web site with the message "Hacked by Chinese." Next, each worm thread checked for the file "c:\notworm" and finding it, the worm would go dormant, which seemed to be a mechanism to limit the worm spread. Finally, each worm thread checked the date. If it was past the 20th of the month, the threads would stop scanning and instead would launch a DoS attack against port 80 of "www.whitehouse.gov." On the 27th of the month, the worm would go dormant permanently although an unpatched machine could be re-infected with a new worm.

A week later of July 19, a second version of Code Red I (Code Red v2) was noticed which spread much faster. Apparently, the programming had been fixed to change the static seed to a random seed, ensuring that the randomly generated IP addresses to probe would be truly random (Berghel, 2000). Code Red v2 worm was able to infect more than 359,000 machines within 14 h (Staniford 2001). At its peak, 2000 hosts were infected every minute. By design, Code Red I stopped spreading itself on July 20.

On August 4, 2000, a new worm called Code Red II (carrying a different, more dangerous payload than Code Red I but self-named "Code Red II" in its source code) was observed exploiting the same buffer overflow vulnerability in IIS web servers (CERT incident note IN-2001-09). After infecting a host, it laid dormant for 1–2 days and then rebooted the machine. After rebooting, the worm activated 300 threads (but 600 for Chinese language systems) to probe other machines to propagate. It generated random IP addresses but they are not completely random; about one out of eight are completely random; four out of eight addresses are within the same class A range of the infected host's address; and three out of eight addresses are within the same class B range of the infected host's address. The attack code ran for 24 hours (but 48 hours for Chinese language systems). The enormous number of parallel threads created a flood of scans, in effect a DoS attack. Then it would write a Trojan horse "explorer.-exe" into the root directories of the hard drives, and cause a reboot to load the Trojan. The Trojan did several things to conceal itself and open a backdoor.

On September 18, 2001, the Nimda worm raised new alarms by using five different ways to spread and carrying a dangerous payload. The fast spreading had a side effect of traffic congestion. Initially, many sites noticed a substantial increase in traffic on port 80. Nimda spread to 450,000 hosts within the first 12 hours. Although none of the infection vectors was new, the combination of so many vectors in one worm seemed to signal a new level of complexity not seen before (CERT advisory CA-2001-26).

- It sent itself by e-mail with random subjects and an attachment named "readme.exe." It finds e-mail addresses from the computer's web cache and default MAPI mailbox. The worm carries its own SMTP engine. If the target system supports the Automatic Execution of Embedded MIME types, the attached worm will be automatically executed and infect the target. Infected e-mail is resent every 10 days.
- It infected Microsoft IIS web servers, selected at random, through a buffer overflow attack called a Unicode Web Traversal Exploit (known for a year prior). This vulnerability allows submission of a malformed URL to gain access to folders and files on the server's drive, and modify data or run code on the server.
- It copied itself across open network shares. On an infected server, the worm writes MIME-encoded copies of itself to every directory, including network shares. It creates Trojan horse versions of legitimate programs by prepending itself to .EXE files.
- It added Javascript to web pages to infect any web browsers. If it finds a web content directory, it adds

a small piece of javascript code to each .html, .htm, or .asp file. This javascript allows the worm to spread to any browser listing these pages and automatically executing the downloaded code.
- It looked for backdoors left by previous Code Red II and Sadmind worms.

Upon infection, it takes several steps to conceal its presence. Even if found, the worm was very difficult to remove because it makes numerous changes to Registry and System files. It creates an administrative share on the C drive, and creates a guest account in the administrator group allowing anyone to remote login as guest with a blank password.

Beginning with the Klez and Bugbear worms in October 2001, "armored" worms contain special code designed to disable antivirus software using a list of keywords to scan memory to recognize and stop antivirus processes and scan hard drives to delete associated files (F-Secure Security Information center; Sophos). Other recent examples of armored worms include Winevar (November 2002) and Lirva (January 2003). More destructively, the Lirva worm, named after the singer, Avril Lavigne, will e-mail cached Windows dial-up networking passwords to the virus writer, and e-mail randon .TXT and .DOC files to various addresses (Symantec Security Response, a). It will connect to a web site "web.host.kz" to download Back Orifice giving complete control to a remote hacker.

In March 2002, Gibe spread as an attachment in an e-mail disguised as a Microsoft security bulletin and patch. The text claimed that the attachment was a Microsoft security patch for Outlook and Internet Explorer. If the attachment is executed, it displays dialog boxes that appear to be patching the system, but a backdoor is secretly installed on the host.

Another trend to dangerous payloads includes installation of keystroke logging software that record everything types on the keyboard, usually recorded to a file that can be fetched by a remote hacker later. The Bugbear, Lirva, Badtrans (November 2001), and Fizzer (May 2003) worms all install a keystroke logging Trojan horse.

The SQL Slammer/Sapphire worm appeared in January 2003, exploiting a buffer overflow vulnerability Microsoft SQL Server announced by Microsoft in July 2002 (with a patch) (CERT advisory CA-2003-04). It is much simpler than previous worms and fits in a 376-byte payload of a single UDP packet. In contrast, Code Red was about 4000 bytes and Nimda was 60,000 bytes. The sole objective seems to be replication due to the absence of a payload. It appeared to test the concept that a small, simple worm could spread very quickly. The spreading rate was surprisingly fast, reportedly infecting 90% of vulnerable hosts within 10 min (about 120,000 servers). In the first minute, the inflection doubled every 8.5 sec, and hit a peak scanning rate of 55,000,000 scans/sec after only 3 min. In comparison, Code Red infection doubled in 37 min (slower but infected more machines). The probing is fast because infected computers simply generate UDP packets carrying the worm at the maximum rate of the machine.

6. CURRENT DEVELOPMENTS

The week beginning on August 12, 2003, has been called the worst week for worms in history, seeing the fast-spreading Blaster, Welchia (or Nachi), and Sobig.F worms in quick succession. Blaster or LovSan arrived first, targeted to a Windows DCOM RPC (distributed component object model remote procedure call) vulnerability announced only a month earlier on July 16, 2003 (CERT advisory CA-2003-20). The worm probes for a DCOM interface with RPC listening on TCP port 135 on Windows XP and Windows 2000 PCs. Through a buffer overflow attack, the worm causes the target machine to start a remote shell on port 4444 and send a notification to the attacking machine on UDP port 69. A tftp (trival file transfer protocol) "get" command is then sent to port 4444, causing the target machine to fetch a copy of the worm as the file MSBLAST.EXE. In addition to a message against Microsoft, the worm payload carries a DoS agent (using TCP SYN flood) targeted to the Microsoft Web site "windowsupdate.com" on August 16, 2003 (which was easily averted). Although Blaster has reportedly infected about 400,000 systems, experts reported that the worm did not achieve near its potential spreading rate due to novice programming.

Six days later on August 18, the apparently well-intended Welchia or Nachi worm spread by exploiting the same RPC DCOM vulnerability as Blaster. It attempted to remove Blaster from infected computers by downloading a security patch from a Microsoft Web site to repair the RPC DCOM vulnerability. Unfortunately, its probing resulted in serious congestion on some networks, such as Air Canda's check-in system and the US Navy and Marine Corps computers.

The very fast Sobig.F worm appeared on the very next day, August 19, only seven days after Blaster (Symantec Security Response, b). The original Sobig.A version was discovered in January 2003, and apparently underwent a series of revisions until the most successful Sobig.F variant. Similar to earlier variants, Sobig.F spreads among Windows machines by e-mail with various subject lines and attachment names, using its own SMTP engine. The worm size is about 73 kilobytes with a few bytes of garbage attached to the end to evade antivirus scanners. It works well because it grabs e-mail addresses from a variety of different types of files on the infected computer and secretly e-mails itself to all of them, pretending to be sent from one of the addresses. At its peak, Sobig.F accounted for one in every 17 messages, and reportedly produced over 1 million copies of itself within the first 24 hr. Interestingly, the worm is programmed to stop spreading on September 10, 2003, which suggests that the worm was intended as a proof-of-concept. This is supported by the absence of a destructive payload, although the worm is programmed with the capability to download and execute arbitrary files to infected computers. The downloading is triggered on specific times and weekdays, which are obtained via one of several NTP servers. The worm sends a UDP probe to port 8998 on one of several preprogrammed servers which responds with a URL for the worm to download. The worm also starts to listen on UDP ports 995–999 for incoming messages, presumably instructions from the creator.

If these recent worm incidents are viewed as proof-of-concept demonstrations, we might draw the following observations:

- Blaster suggests a trend that the time between discovery of a vulnerability and the appearance of worm to exploit it is shrinking (to one month in the case if Blaster).
- Blaster demonstrates that the vast majority of Windows PCs are vulnerable to new outbreaks.
- Sobig shows that worm writers are creating and trying out successive variants in a process similar to software beta testing, and if variants are being written by different people, it suggests an increasing degree of coordination among worm writers.

One observation is certainly evident from recent worm incidents. Despite firewalls, intrusion detection systems, and other network security equipment, worm outbreaks continue to spread successfully. No one doubts that new worm outbreaks can be expected in the future. Indeed, outbreaks have become so commonplace that most organizations have come to view them as a routine cost of operation.

The reasons for the continued state of vulnerability are complex. The problem is seen by some at a simplistic level as a struggle between virus writers and the antivirus industry. However, the problem is much larger, involving the entire computer industry. Worms and viruses will continue to be successful as long as computers have security vulnerabilities that can be exploited. These vulnerabilities continue to exist for number of reasons. First, software has become enormously complex, and perfectly secure software has become more difficult. For example, buffer overflow attacks have been widely known since 1995 but this type of vulnerability continues to be found often (on every operating system). Hence, the root problem is that unsecure software continue to be written. Second, when vulnerabilities are announced with corresponding software patches, many people are unaware of them or slow to apply patches to their computer for various practical reasons. Frequent patching takes significant time and effort, and faith that the patches will not cause troublesome side effects. Third, many people are unaware or choose to ignore the security problems. This problem is particularly relevant for home PC owners who often must be educated about security and then make the effort to acquire antivirus software, personal firewalls, or other means of protection.

In addition to technical causes, a sociological reason is the lack of accountability for worm writers. The long-held general perception has been that worms and viruses are low risk crimes. It is notoriously difficult to trace a virus or worm to its creator from analysis of the code, unless there are inadvertent clues left in the code. Even if a virus creator is identified and arrested, cases are difficult to prosecute and prove malicious intention (not just recklessness). Historically, long prison sentences have been perceived as overly harsh for convicted virus writers who have tended to be teenagers and university students. Therefore, sentences have tended to be light. The author of the 1988 Internet worm, Robert Morris, was sentenced to three years of probation, 400 hours of community service, and a $10,000 fine. Chen Ing-hau was arrested in Taiwan for the 1998 Chernobyl virus but released when no official complaint was filed. Onel de Guzman was arrested for writing the 2000 LoveLetter virus resulting in $7 billion of damages, but he was released due to the lack of relevant laws in the Philippines. Jan De Wit was sentenced for the 2001 Anna Kournikova virus to 150 hr of community service. David L. Smith, creator of the 1999 Melissa that caused at least $80 million in damages, was sentenced to 20 months of custodial service and a $7500 fine. In the absence of a serious legal deterrent, the general perception persists that virus writers can easily avoid the legal consequences of their actions.

7. CONCLUSIONS

This review of the evolutionary development of viruses and worms leads to a number of conclusions about our current situation. First, new virus and worm outbreaks may be expected for the foreseeable future. Software will continue to have vulnerabilities that will be expoited by virus writers. Software patching will continue to be problematic for various practical reasons. Second, new worms may appear sooner after a vulnerability is discovered, increasing their opportunity to spread quickly through a largely unprotected population. Third, new worms have been successfully exploring new infection vectors (e.g., peer-to-peer networks, instant messaging, wireless devices). This also increases the chances that a new outbreak may be successful. Fourth, future worms will likely see a rapid succession of variants, again increasing the chances that at least one variant will spread successfully.

The general conclusion is that a new fast-spreading worm outbreak is a real possibility. Continuing recent trends, a new worm may be able to saturate the vulnerable population within minutes or possibly seconds. This points to an urgent need for better, automated means of protection. This protection could take the form of a global coordinated antivirus system. This system would need the capabilities to reliably and quickly detect signs of a new worm outbreak anywhere in the world; broadcast notifications of a new worm throughout the system; and automatically quarantine the outbreak by selectively isolating the infected computers from the rest of the Internet.

REFERENCES

Cohen, F. (1987). Computer viruses: theory and experiments. *Computers and Security* 6:22–35.

Cohen, F. (1994). *A Short Course on Computer Viruses*. 2. NY: John Wiley & Sons.

Grimes, R. (2001). *Malicious Mobile Code: Virus Protection for Windows*. Sebastopol, CA: O'Reilly & Associates.

Ludwig, M. (1998). *The Giant Black Book of Computer Viruses*. Show Low, AZ: American Eagle Publications.

Harley, D., Slade, R., Gattiker, R. (2001). *Viruses Revealed*. NY: Osborne/McGraw-Hill.

von Neumann, J., Burks, A. (1966). *Theory of Self-Reproducing Automata*. Urbana, IL: University of Illinois Press.

Shoch, J., Hupp, J. (1982). The 'worm' programs — early experience with a distributed computation. *Communications of ACM* 25:172–180.

Spafford, E. (1989). The Internet worm program: an analysis. *ACM Computing and Communication Review* 19:17–57.

Smith, G. (1994). The Virus Creation Labs: A Journey into the Underground. Tucson, AZ: American Eagle Publications.

CERT advisory CA-2001–03. VBS/OnTheFly (Anna Kournikova) malicious code. http://www.cert.org/advisories/CA-2001-03.html.

CERT incident note CA-1999–02. Happy99.exe Trojan horse. http://www.cert.org/incident_notes/IN-99-02.html.

Cass, S. (2001). Anatomy of malice. *IEEE Spectrum* 38:56–60.

CERT advisory CA-1999-04. Melissa macro virus. http://www.cert.org/advisories/CA-1999-04.html..

CERT advisory CA-1999-06. ExploreZip trojan horse program. http://www.cert.org/advisories/CA-1999-06.html.

CERT advisory CA-2000-04. Love letter worm. http://www.cert.org/advisories/CA-2000-04.html.

CERT incident note IN-2001-02. Open mail relays used to deliver Hybris worm. http://www.cert.org/incident_notes/IN-2001-02.html.

CERT advisory CA-2001-11. Sadmind/IIS worm. http://www.cert.org/advisories/CA-2001-11.html.

CERT incident note IN-2001-08. Code Red worm exploiting buffer overflow in IIS indexing service DLL. http://www.cert.org/incident_notes/IN-2001-08.html.

Berghel, H. (2001). The Code Red worm. *Communications of ACM* 44:15–19.

Staniford, S. (2001). Analysis of spread of July infestation of the Code Red worm. http://www.silicondefense.com/cr/july.html.

CERT incident note IN-2001-09. Code RedII: another worm exploiting buffer overflow in IIS indexing service DLL. http://www.cert.org/incident_notes/IN-2001-09.html.

CERT advisory CA-2001-26, Nimda worm. http://www.cert.org/advisories/CA-2001-26.html.

F-Secure Security Information Center. F-Secure virus descriptions: Klez. (2001). http://www.europe.f-secure.com/v-descs/klez.shtml.

Sophos. (2001). W32/Bugbear-A. http://www.sophos.com/virusinfo/analyses/w32bugbeara.html.

Symantec Security Response. a. W32.lirva.C@mm. (2003). http://securityresponse.symantec.com/avcenter/venc/data/w32.lirva.c@mm.html.

CERT advisory CA-2003-04. MS-SQL server worm. http://www.cert.org/advisories/CA-2003-04.html.

Moore, D., Paxson, V., Savage, S., Shannon, C., Staniford, S., Weaver, N. The spread of the Sapphire/Slammer worm. http://www.caida.org/outreach/papers/2003/sapphire/sapphire.html.

CERT advisory CA-2003-20. W32/Blaster worm. Aug. 11, 2003. http://www.cert.org/advisories/CA-2003-20.html.

Symantec Security Response, b. W32.Sobig.F@mm. (2003). http://securityresponse.symantec.com/avcenter/venc/data/w32.sobig.f@mm.html.

17

Infrastructure Web: Distributed Monitoring and Managing Critical Infrastructures

GUOFEI JIANG, GEORGE CYBENKO, and DENNIS McGRATH
Institute for Security Technology Studies, Thayer School of Engineering, Dartmouth College, Hanover, NH, USA

ABSTRACT

National-scale critical infrastructure protection depends on many processes: intelligence gathering, analysis, interdiction, detection, response and recovery, to name a few. These processes are typically carried out by different individuals, agencies and industry sectors. Many new threats to national infrastructure are arising from the complex couplings that exist between advanced information technologies (telecommunications and internet), physical components (utilities), human services (health, law enforcement, emergency management) and commerce (financial services, logistics). Those threats arise and evolve at a rate governed by human intelligence and innovation, on “internet time” so to speak. The processes for infrastructure protection must operate on the same time-scale to be effective. To achieve this, a new approach to integrating, coordinating, and managing infrastructure protection must be deployed. To this end, we have designed an underlying web-like architecture that will serve as a platform for the decentralized monitoring and management of national critical infrastructures.

1. INTRODUCTION

Modern threats to critical national infrastructure are evolving at the same rate as the technology on which that infrastructure is based. This is a key axiom of the work described in this paper. To illustrate the point, consider the following chronology of events related to the recent distributed denial of service (DDOS)* attacks launched against major e-commerce companies

Early summer of 1999	DDOS capabilities are demonstrated at a European “hacker festival.”
Last summer of 1999	First DDOS attacks at the University of Minnesota are detected and documented.
November 1999	A workshop on DDOS attacks and defense mechanisms is hosted by the Computer Emergency Response Team (CERT),[a] Carnegie Mellon University.
December 1999	Programs for detecting DDOS “zombies” are distributed.
February 2000	DDOS attacks are launched against major internet sites.
March 2000	The possibility of a DDOS-type attacks against the 911 system is identified.
April 2000	DDOS-type attacks against the 911 system are suspected in Texas.

[a]http://www.cert.org

*http://www.sans.org/ddos_roadmap.htm

Sometime in the future DDOS attacks within the financial sector, using automatically generated consumer trading, are detected.

This chronology illustrates three major points:

a. The time intervals between when a new threat is identified, when it manifests itself and when it is modified (mutated) into different forms are relatively short and appear to be shrinking.
b. Cyber security and physical security are rapidly becoming intertwined, and attacks against information systems can have immediate and profound effects on the physical systems that depend on them.
c. Threats within one sector (telecommunications/internet) can easily spill over into other sectors such as human services (the 911 system) and the financial system.

To meet these challenges, we need to leverage modern information technologies and create an infrastructure protection process that can operate seamlessly at an accelerated timescale. Moreover, that process must be able to monitor and manage the complex interactions between infrastructure segments that are becoming the norm. This is especially important considering the fact that many recent attacks on national infrastructure have been credited the pranksters and individuals working alone or in small groups. We have not yet really seen what kinds of damage well-financed, coordinated, professional attacks are capable of creating.

Like the World Wide Web, the infrastructure web should have the following characteristics:

1. It should be decentralized, asynchronous, and redundant.
2. New elements can be added to it or old elements can be removed from it by authorized personnel but without centralized control.
3. It should be searchable and self-organizing.
4. It should allow new services to be built easily on top of existing services.
5. It should allow for multiple, redundant communication paths between entities.

Section 2 of this paper describes the various stages in the Critical Infrastructure Protection process today together with our vision for how those stages can be integrated. Special attention is given to infrastructure related to information technology, namely internet and telecommunications, but we indicate how the ideas can be generalized to other infrastructure segments. Section 3 describes the conceptual organization of the infrastructure web that we are currently implementing. The functional operation is illustrated through some examples. Meanwhile Section 3 also gives a brief technical description for how the infrastructure web can be implemented, using current computing and networking technologies.

Section 4 discusses some related work, and Section 5 is a summary.

2. THE INFRASTRUCTURE PROTECTION PROCESS

The emergency management, public health and more recently computer security communities have decomposed their management processes into smaller, logically concise stages. For example, the DARPA Information Assurance program is using the three-stage "Protect–Detect–React" paradigm to organize work within that area.[†] Figure 1 shows the six stages we propose for Information Infrastructure Protection. These stages roughly correspond to stages used in other emergency management areas with different degrees of granularity perhaps. We briefly describe each stage and its relationships with other stages.

2.1. Intelligence

The first step in infrastructure management is intelligence gathering about emerging threats. This is typically done using human intelligence reporting, analysis of unusual incidents, and information harvesting from open sources

[†]http://www.darpa.mil/iso/ia/

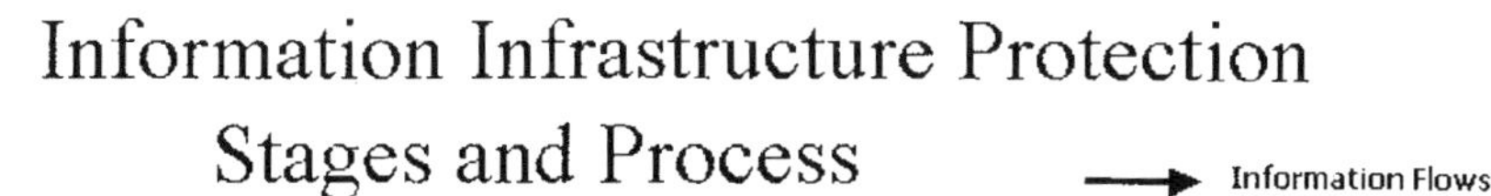

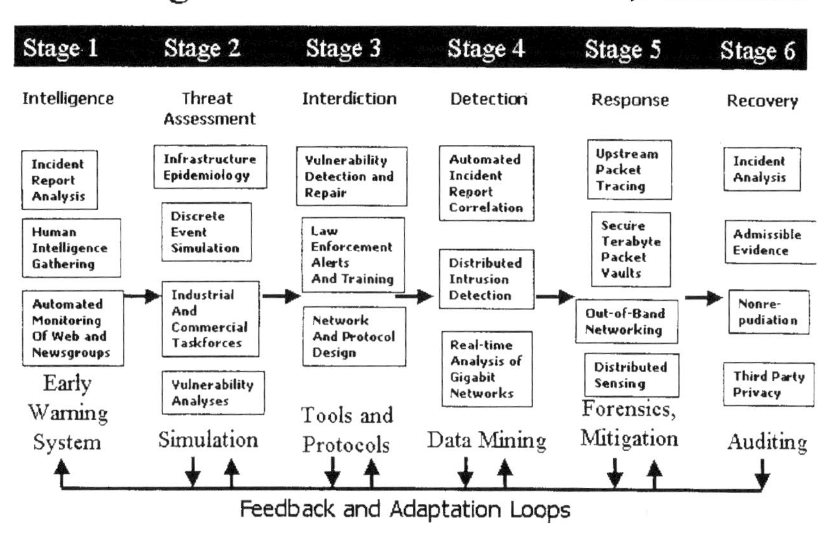

Figure 1 The information infrastructure protection process.

such as the web and news sources. This is the early warning system that can identify new threats early in the process, before they manifest themselves in real attacks or disasters. "Red teaming", namely the use of selected experts for scenario building and threat design for proactive analysis, is an important part of this stage. We include that in the "human intelligence" component.

Figure 2 identifies three sources of intelligence for early threat identification: incident analysis, human intelligence, and automated tools for harvesting and organizing information from open sources such as the web and newsgroups. In the information infrastructure protection problem, early evidence of threats is often proposed and discussed in such open sources. Such open sources are useful for human-initiated threats but not so useful perhaps for predicting complex interactions between infrastructure elements or natural events and design flaws. Human intelligence is more important for identifying those threats.

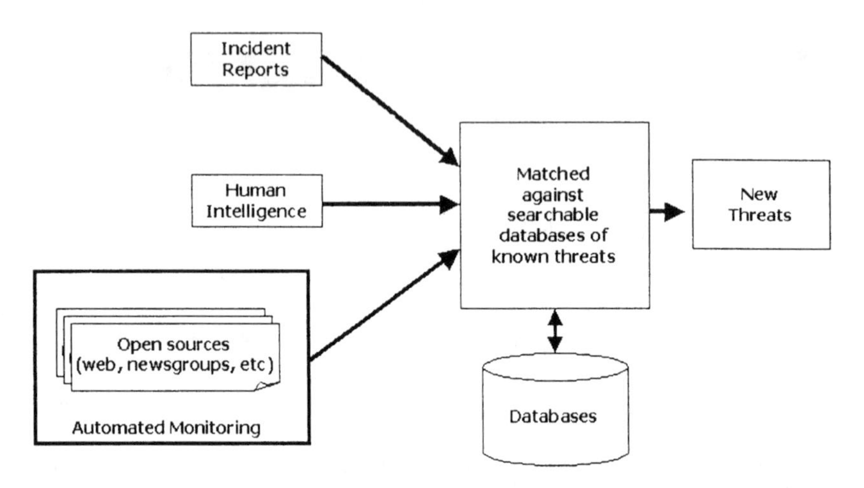

Figure 2 Early threat detection and warning.

From the point of view of automating this stage of the process, automated incident report analysis and monitoring of open web- and internet-based sources are most promising. Several organizations already provide on-line access to incident reports and threat alerts (see http://www.cert.org for example) although those resources are not organized to allow powerful search capabilities through a database engine interface.

Ideally, a new incident report could be quickly and automatically matched against an on-line database of previously seen threats and attacks to see if the threat is novel is known. Today, this stage of early warning is done by experts who rely on their own memories, networks of colleagues and ad hoc searches of archives of previous attacks.

Automated monitoring of the web and various news groups for early threat identification is technically possible today[‡] but not done to our knowledge. We are currently developing such a capability.

[‡]http://informant.dartmouth.edu.

2.2. Threat Assessment

Once a new threat is identified, risk assessment and some sort of "cost-benefit" analysis of responding to the threat must be performed. This stage requires some sort of epidemiological model of how an attack or failure based on the threat will manifest itself and how it will affect other infrastructure systems. Basically, the question is: what its dynamics are? Related to this of course are the costs associated with containment or interdiction vs. the costs of an attack or failure based on that threat. As is often the case in defensive strategies, the cost of defense can be much higher than the cost of the attack, but that must be weighted against the social and human cost of major systems failures.

At present, our understanding of "infrastructure epidemiology" is very poor, at least in the open literature. The challenge here is to develop quantitative models of how vulnerabilities are distributed nationally or even globally, and how failures based on those vulnerabilities can cascade through the overall infrastructure.

Such analyses will probably have to rely on large-scale discrete event simulations since closed-form solutions are highly unlikely. Government, industrial, and commercial task forces must be able to provide quick and reliable input into the vulnerability assessment process so that some form of realistic cost-benefit analysis can be performed in the threat assessment stage (Figure 3).

2.3. Interdiction

The interdiction stage of infrastructure protection attempts to proactively prevent or prohibit attacks or failures, based on known threats. Virus scanners, software patches, and improved network designs and protocols are examples of interdiction in the information infrastructure segment. An important element of interdiction is the training of system operators and law enforcement personnel, especially at the state and local levels since these communities are typically the first responders to attacks and failures.

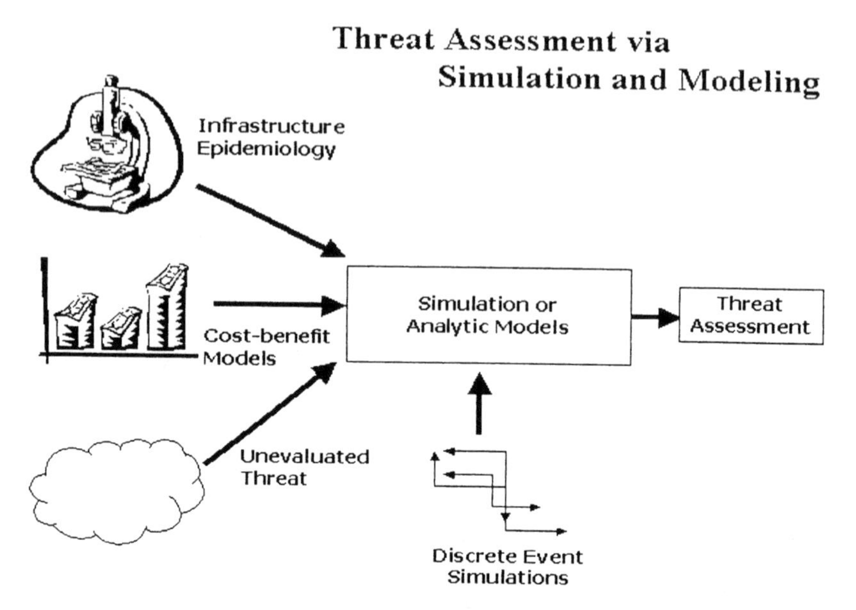

Figure 3 The threat assessment stage.

These ingredients in the interdiction stage typically operate at different timescales. For example, the deployment of more robust and secure designs and protocols can take many years to permeate the infrastructure because of lock-in effects. On the other hand, software patches and virus scanning updates occur on the timescale of weeks quite often. The training of early responders such as system operators and law enforcement and emergency management personnel is problematic because of the huge demands on time and expertise in those sectors. The rate at which new threats and vulnerabilities are arising outstrips the ability of such personnel to attend training meetings and courses so that remotely accessible, distance training using networked interactive material is necessary. Cost-benefit analysis is essential to identify threats and vulnerabilities that are most likely to have high impact because existing time commitments and obligations preclude the ability of first responders to be prepared for all possible failures. This stage of the process must focus on interdiction in high-cost and/or high-probability events.

2.4. Detection

The detection of actual failures or attacks is enabled by monitoring distributed "sensors" that are positioned throughout the infrastructure itself. Raw sensor data must be harvested, mined, correlated, and otherwise analyzed. Examples of such sensors include computer network monitors (based for example on SNMP agents or packet analyzers), public health records, medical laboratory results, environmental monitoring stations, financial market trend monitors, and so on. Human observations in the form of natural language reports are also relevant to this stage.

Whereas the Early Warning System part of the process is meant to anticipate attacks and failures through proactive intelligence gathering and analysis, this stage is meant to respond to mature attacks and imminent failures. Ideally, threat assessment and interdiction have prepared the community for these events but may not always be the case.

The challenge in automating this stage of the process lies in flagging anomalous events that have not been seen before without generating large numbers of false positives. This requires training an automated system on "normal" and known behaviors, flagging behaviors that fall outside this regime. The technical challenge here is that many new behaviors emerge in the course of natural, non-threatening operating modes. Much work remains to be done in this area.

2.5. Response

Once an attack or failure has been detected, an appropriate response is required. We focus on law enforcement or internal auditing responses to information infrastructure events. A major challenge in responding to cybercrime and cyberterrorism attacks is identifying the source of the problem. This requires forensic techniques that allow building a trail of legal evidence for future investigation while respecting the privacy of third parties. These considerations require the ability to do fast and reliable upstream packet tracing, something that currently require time consuming and relatively slow operator intervention. Moreover, the fact that many internet links are now operating in the multiple megabit and even gigabit per second range, archiving network traffic for forensic analysis is a major technical challenge. Early work in this area is promising but much development remains to be done.

Another fundamental challenge is responding to infrastructure failures and attacks is that the very system, namely the telecommunications networks, that responders will have to use to coordinate a response are themselves part of the infrastructure and highly vulnerable to failures. Any future infrastructure web architecture must provide for "out-of-band" and otherwise redundant communication capability.

This can be accomplished through the use of multiple communication channels based on different protocols implemented by different vendors so that a single vulnerability does not compromise the whole system. In this case, standardization is bad for survival and we need heterogeneous systems. Additional out-of-band communication capability can be achieved by radio and satellite networking which is currently being investigated on several fronts (Figure 4).

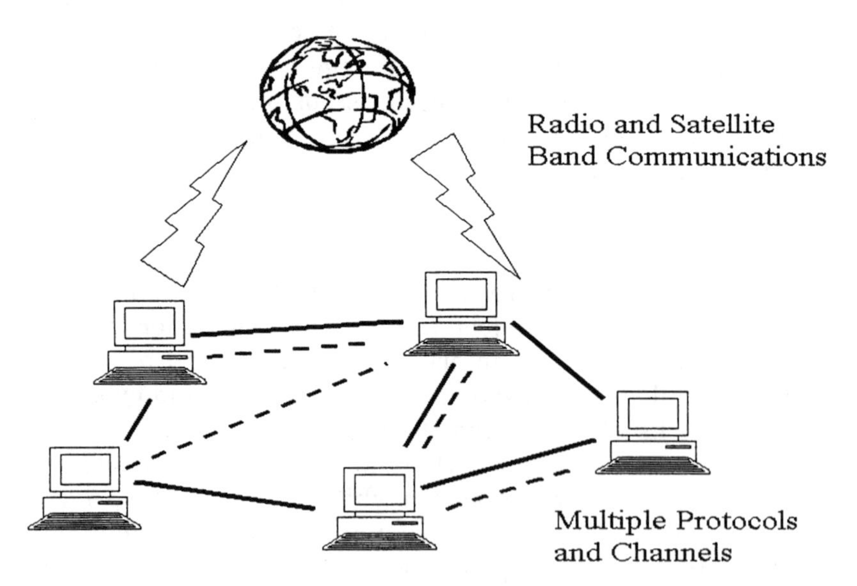

Figure 4 Out-of-band and redundant communications channels.

2.6. Recovery

In the law enforcement arena, recovery from an attack or other criminal activity related to national infrastructure includes archiving non-reputable evidence without violating privacy laws and standards. Complete analysis of the incident is required to learn from it and to archive its characteristics in appropriate databases for future use in detection and training. Technical challenges here include training of first responders on the appropriate forensic techniques that accomplish these goals.

3. ARCHITECTURE

According to the report of the President's Commission on Critical Infrastructure Protection,[§] the infrastructure networks of greatest importance to national security and stability include telecommunications, electrical power systems, gas and oil, banking and finance, transportation, water supply systems, government services, and emergency services. To effectively protect these critical infrastructures, it will be necessary to have a system in place that can monitor and manage these very large, complex and dynamic networks. Our proposed infrastructure web system provides such a basic architecture, as well as the underlying paradigms by which a problem of this scale and scope can be addressed.

The above section has discussed the various stages in the critical infrastructure protection process and our vision for how those stages can be integrated. So now the question is: how to integrate and implement these stages and visions into a real monitoring and management system? We propose that national infrastructure web networks be built up with four types of basic distributed components: Directory service, Infrastructure server, Sensor web and Emergency Information Search server. All these distributed components will be organized and integrated throughout the national wide networks with Sun's Jini system (Edwards, 1999). Jini is

---

[§]http://www.ciao.gov

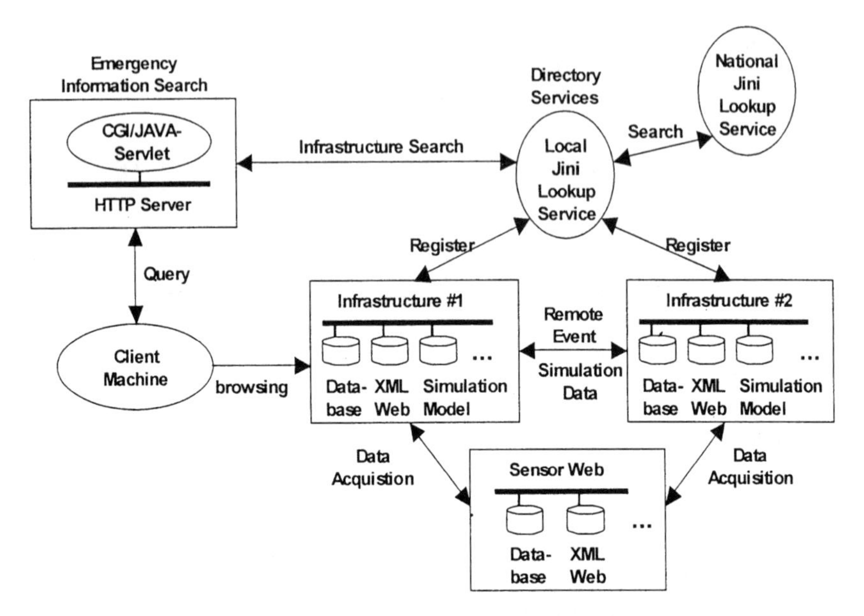

Figure 5 The architecture of the Infrastructure web.

designed for deploying and using services in a network and enables the construction of dynamic, flexible, and robust systems from independent distributed components. Further, access to Jini's source code has fostered a thriving community of developers who continue to enhance and expand Jini's capability. A framework of the infrastructure web architecture is shown in Figure 5. With this kind of architecture, we believe that the infrastructure web system can be exploited as a platform to implement our distributed infrastructure assurance vision.

Currently several efforts are underway in US to build the computational grids, such as Globus (Foster and Kesselman, 1998) and DARPA CoABS.[¶] These projects are developing the fundamental technology that is needed to build the computational grids, execution environments that enable an application to integrate geographically distributed instruments, sensors, data products, displays, and computational

[¶]http://coabs.globalinfotek.com.

and information resources. Here so-called grid computing refers to computing in a distributed networked environment in which computing and data resources are located throughout the network. The vision is that these grid infrastructures will connect multiple regional and national computing power and data resources that support dramatically new classes of applications. In this sense, the infrastructure web system is actually a national wide grid that is specially developed for the critical infrastructure protection. We believe that with more and more sensors and other services on the infrastructure servers come to online, the infrastructure protection grid will grow quickly. Currently we are investigating the possible standard and transparent data exchange format, communication protocol, user interfaces and APIs (application programming interface) among these infrastructures. So later by the integration with these existing distributed services, some new applications or services can be easily implemented for infrastructures' protection and then serve as new "existing services" for the further development of other applications or services. As the result, the scale of the grid is growing.

3.1. Infrastructure Server

In the infrastructure web system, one infrastructure network server represents one critical infrastructure element (such as a hospital or power plant) in the physical world, and the server's IP address is the unique identification for the infrastructure elements. Basically, the infrastructure server will have a real-time database, an XML-based web, a simulation model and other infrastructure protection services running on various ports of a single server. The database acquires real-time data from the sensor web or other sources (e.g., host-based detection systems). These data consist of the security status, internal state information, and so on. Some data will be displayed on the web in real time to show the infrastructure's current status, and some will be used as input to simulations for automated evaluation (such as the threat assessment). Currently we employ a Mysql database server in our implementation because it is a free open source database server, and it is supported on multiple platforms. Open Database Connectivity (ODBC) and Java Database Connectivity (JDBC) are two of the most popular database interfaces that make database access transparent. Applications written in modern programming languages, such as, Java, Visual Basic and C++, can easily use the ODBC or JDBC function from the database driver to interact with remote databases. This kind of open database connectivity can be used as a standard data exchange approach between the distributed applications or services. On every server's web pages, the relational database server type and the table layout will be described, so a remote client can download the driver for that type and use ODBC or JDBC functions to call an SQL query to access the database. The advantage of this data exchange approach is that specific knowledge of the various communication protocols and programming interfaces is not needed for the client. Once the remote client knows the table layout for the database, he knows how to access the data. Moreover, the security policy of the database server uses multi-level security and authorizes remote clients different data access permissions.

It is widely accepted that extensible markup language (XML) is the most promising Internet technology development since Java. XML is a license-free, platform-independent language for describing and publishing structured data in text form. Where HTML only allows users to see formatted text, XML allows users to understand and use self-describing and structured data. For each of the eight infrastructure categories listed in the President's Commission on Critical Infrastructure Protection, an Infrastructure Markup Language (IML), created with XML document type definitions (DTDs), will clearly describe the attributes of the category in standard formats. XML is also an excellent vehicle for the interchange of data among different applications. By browsing an infrastructure server's XML pages, clients can check on or many state variables of the critical node, such as the packet traffic throughput of an important LAN infrastructure. The relationships between related infrastructures will be described by XML's X-Links and X-pointers (Harold, 1999). Meanwhile other web technologies such as Java applets and JavaScript will also be used to describe the infrastructures.

As essential part of the infrastructure server consists of appropriate analytical and simulation models of the infrastructures together with a description of their behavior under dynamically changing interconnections. Just as pins are wired in chips on a PCB (printed circuit board), large-scale discrete-event simulations can be implemented for threat assessments by "wiring" infrastructure models' input and output. We believe that with some adaptive learning technologies such as neurocomputing and evolutionary computing (Bertsekas and Tsitsiklis, 1996) improved planning, control, and coordination strategies can be derived with simulations. These strategies and policies will help these related infrastructures cooperate efficiently once a disaster or attack occurs. Other infrastructure protection applications and services will be implemented on the infrastructure server based on the specific requirements of the infrastructure element category. For example, the early warning system can be implemented as an information infrastructure in our architecture to offer automated intelligence collection services. But the basic development process for these applications will be similar: search and collect the distributed sensor and infrastructure data that are related to that infrastructure; develop the application program to process these data and then to verify whether the infrastructure is running well or under threat (which needs specific knowledge support from experts in that area); if not, take steps in the application programs to respond and recover the possible failure. Meanwhile, if the application is going to offer data to other applications, it should also report the data with the standard user interface and API as we discussed.

3.2. Directory Service

The infrastructure web will have two directory service levels: a local/state level directory service and a national level directory service. All of these directory services will be implemented with Jini lookup services. Infrastructure elements need to register themselves in the local Jini lookup services, and the local lookup services will register themselves in the national level lookup services. Currently we are investigating the architecture of this distributed directory services system in order to make it robust, reliable, and efficient. For example, should the system be organized with the hierarchical structure like the domain name service (DNS) or the peer to peer structure like the Gnutella?[1] Each infrastructure element registers with attributes such as category, location, the server IP and URL, proxy interface program, and so on. Jini's attribute mechanisms support both type-based and content-based search styles, and make searching for particular attributes simple, quick, and effective. Standard taxonomies and specifications will describe infrastructures and their services accurately, but those taxonomies and specifications do not yet exist.

A typical system implemented with Jini has five basic concepts: Discovery, Lookup, Leasing, Remote Events, and Transactions. Jini's ability to support spontaneously created, self-healing communities of distributed components is based on these concepts, and Java's Remote Method Invocation (RMI) and object serialization (Weber, 1998) techniques make the implementations of these concepts possible. Since the infrastructure web will be organized and integrated with Jini system, it will inherit these features from Jini. For example, with Jini's leasing concept, all infrastructures need to "sign" a lease with the lookup service in the registrations. Once the lease expires, the infrastructure will automatically be removed from the lookup service. In this way, a Jini lookup service has a self-healing ability for its directory management, and clients will not receive outdated information. The relationship between related infrastructures can be described using the remote event concept. For example, infrastructure no. 1 can tell a related infrastructure no. 2

[1] http://www.gnutella.wego.com.

which status message it cares about, the infrastructure no. 2. will be automatically notified of changes by remote events from infrastructure no. 2. In this way, geographically distributed infrastructure elements can cooperate efficiently to detect, respond, and recover from the intrusion and attack. Using these Jini concepts, we believe that the infrastructure web can be easily implemented with the required characteristics that are proposed in Section 1.

3.3. Sensor Web

The ability to monitor the states of distributed infrastructure elements is another essential part of our system. In the analysis and detection of DDoS attacks, an analyst or upstream packet tracing system needs packet information from local machines as well as remote routers or firewalls. Here our sensor web system is actually a large-scale Distributed Smart Sensor Network (DSSN) that collects distributed sensor information both from intelligent software sensors as well as smart hardware sensors. The sensor web registers its sensors with the Directory Service, and the sensors provide distributed data sensing services to clients. Examples of sensors include computer network monitors (based for example on SNMP agents or packet analyzers), public health records, medical laboratory results, environmental monitoring stations, financial market trend monitors, and so on. Human observations in the form of natural language reports are also relevant to this stage. Like the infrastructure servers, every sensor web system has an XML-web and a database server included. The XML-web is used to describe detailed sensor information and report real-time sensor data on the web pages. After the sensor web system acquires the data from its sensors, it writes the data into its database periodically. All other remote applications or services can use ODBC or JDBC to access the data.

The advances in measurement devices have reduced cost to the point where it is now viable to develop large-scale distributed sensing systems. Meanwhile the advances in processor technology allow for relatively low-cost, low power, compact distributed processing integrated within these sensor devices, commonly referred to as smart sensors. Intelligent or smart sensors capable of parsing and filtering only the necessary or desired information allow for efficient use of memory, precious wireless bandwidth, and battery power needed for the transfer of sensor information. Before sending sensor information to related infrastructure elements, the sensor web system will preprocess the data from the distributed sensors using methods such as data filtering, data fusion, and data mining. More information about our distributed sensor web systems can be found in Michael (2000).

3.4. Emergency Information Search

When an infrastructure element fails or is attacked intentionally, the damage needs to be assessed for a rapid recovery. In some cases, lost services might be covered by nearby or related infrastructure elements. The status of those other elements should be checked to help the coordinators to make informed decisions. Unfortunately, this kind of online emergency information search and response system does not exist at this time. The infrastructure web would have a nationwide directory service of all critical infrastructure elements and could therefore perform the role of emergency response system.

Just like the "911" telephone emergency systems, the emergency information server should have a special and well-known domain name, and the emergency query forms should be well formatted. After clients submit the query, the HTTP server will transfer the query data to CGI or Java Servlet programs. These programs will process the query data and submit a formatted attributes template to the Jini lookup services. Then Jini systems will search through lookup services and return all relevant infrastructure element information and URL's. Then, using the XML-based web, clients can check the real-time status and internal states of other infrastructure elements that might assist in recovery.

4. RELATED WORK

4.1. Grids Projects

While commercial companies are making enterprise component-based frameworks available on the market such as Microsoft's DCOM (Distributed Component Model) and Sun's EJB (Enterprise JavaBeans), some more ambitious grids projects like Globus and CoABS are under the way for distributed computing. The Globus project is developing the fundamental technology that is needed to build computational grids. Grids are persistent environments that enable software applications to integrate instruments, displays, computational and information resources that are managed by diverse organizations in widespread locations. Meanwhile agents are a new software technology perfectly situated to take advantage of today's computing infrastructure and software development. These DARPA CoABS research community is developing a prototype "agent grid" as an infrastructure for the run-time integration of heterogeneous multi-agent and legacy systems.

While these grids projects are focusing on how to build computational grids, our infrastructure web project intends to build specific grids for distributed infrastructure protection. Because all these grids projects try to use widespread existing computational and informational resources to support dramatically new classes of applications, the concern about framework, architecture, and communication protocol for the projects should be similar.

4.2. UDDI

The UDDI (Universal Description, Discovery and Integration) project is a commercial effort that shares some of the same objectives of the infrastructure web.** The UDDI project is an open industry initiative enabling businesses to discover each other and define how they interact via the Internet and share information in a global registry

**http://www.uddi.org/.

architecture. UDDI aims to be a mechanism that will enable businesses to find and transact with one another via their preferred applications. UDDI is also a framework for Web services integration that contains specifications for service description and discovery. The UDDI initiative is currently led by three companies (Ariba, IBM, and Microsoft), but eventually it will be turned over to a standards organization. The UDDI Business Registry is operated as a Web service supporting the UDDI specifications, and an Operator's Council helps set policy and quality of service issues for operators. The UDDI vision of a global registry is similar to this infrastructure web plan, and we might leverage UDDI technology and/or standards to realize the infrastructure web vision. In the abstract, the plans are similar in that both are methods for connecting disparate elements offering a wide variety of services via flexible lookup services. Where commercial companies seek to conduct business transactions, infrastructure elements will share critical information. Certainly, the standardization effort associated with UDDI can serve as a guide for the difficult task of standardizing the format of published data from independent data providers.

5. SUMMARY

The information revolution has introduced computers and the Internet into every corner of our society. Today we are relying more and more on the computer-controlled systems, but these systems are vulnerable to intrusion and destruction. The recent DDoS attacks against e-commerce companies have raised concerns about how to cope with cybercrime and cyberterrorism. Potential future attacks on critical infrastructure could cripple the nation by denying or disrupting the delivery of critical services. So how can we protect our national critical infrastructures from cyberterrorism? In this paper, we proposed six stages for the information infrastructure protection: intelligence gathering, analysis, interdiction, detection, response and recovery and out vision for how these stages can be integrated, with some detailed discussion on our plans for realization of an infrastructure web. The system will be a platform for decentralized monitoring and managing critical national infrastructures.

We believe that after the system is implemented and deployed, it will function as follows: while sensors web systems collect data for all critical infrastructures, the services on every infrastructure server monitor these data to verify that the infrastructure is running well. Once the infrastructure server detects attacks or failures, related infrastructure elements will be notified, and response and recovery steps will be taken automatically. Meanwhile, an emergency response coordinator can search and check real-time data from all appropriate infrastructures and choose the appropriate pre-simulated control strategy to respond to and recover from the attacks or failures. Success of this vision will depend on more than technology. The integration of the various agencies and organizations will require more complete knowledge of the operations of the various infrastructure categories from different fields. Organizations and domain experts will need to analyze the specific requirements for their areas, and more academic and industrial research on infrastructure protection is needed. Eventually state and local government cooperation and support will also be required to deploy this system.

ACKNOWLEDGEMENTS

This research was supported by the Office of Science and Technology in National Institute of Justice (contract 2000-DT-CX-K001), the National Science Foundation (NSF contract CCR-9813744), Defense Advanced Research Projects Agency (DARPA contract F30602-98-2-0107) and the Air Force Office of Scientific Research (AfoSR contract F49620-97-1-03821). All opinions expressed here are solely those of the authors.

REFERENCES

Edwards, W. K, (1999). *Core Jini*. Perntice Hall.

Foster, I., Kesselman, C. (1998) (1998). *The Grid Blueprint for a New Computing Infrastructure*. San Franscisco: Morgan Kaufmann Publishers.

Harold, E. R. (1999). *XML Bible*. IDG Books Worldwide.

Bertsekas, D. P., Tsitsiklis, J. N. (1996). *Neuro-Dynamic Programming*. Massacussetts: Athena Scientific.

Weber, J. L. (1998). *Using Java 1.2*. Que Publishing.

Michael, G. C., Okino, C. (2000). *A Study of Distributed Smart Sensor Networks*. Dartmouth College, Thayer School of Engineering, Technical Report Preprint, March.

18

Firewalls: an Expert Roundtable

JAMES P. ANDERSON
james P. Anderson Company

SHEILA BRAND
National Security Agency,
US Department of Defense

LI GONG
JavaSoft, Sun Microsystems

THOMAS HAIGH
Secure Computing Corp.

A panel of distinguished experts in computer security—representing firewall vendors, government organizations, consultants, and educators—meet online to discuss firewalls and their uses.

In the 1970s, the US Defense Advanced Research Projects Agency fostered the development of the Arpanet, which began as a modest network for academic and research use. Arpanet has blossomed into today's Internet, which is becoming an essential tool for personal, commercial, and government information sharing. The Internet's increasing popularity has brought with it a downside: more-and-more severe-security risks. With initial funding from the US government, the computer security community has developed products called firewalls that help protect users' systems from harm when they connect to the Internet and other networks beyond their control. Dozens of commercial firewall products on the market worldwide are enjoying healthy sales. At the request of guest editors Deborah Cooper and Charles Pfleeger, the following panel of experts met online to examine the effectiveness of firewalls and suggest improvements and alternatives that can make your system safer:

1. WHAT ARE FIREWALLS?

Firewalls are computers or communication devices that restrict the flow of information between two networks. Typically, firewalls are configured to protect one or more machines inside a protected domain. Erected between an organizational network and the Internet or any other outside network, a firewall is intended to prevent a malicious attacker who has control of computers outside the organizations's walls from gaining a foothold inside. Used within an organization, a firewall can limit the amount of damage from an intruder who does penetrate the organization's internal network. An intruder may break into one set of machines but the firewall can protect others, perhaps containing more sensitive data or computations.

A firewall implements a security policy, the set of rules that define what kinds of interactions are allowed between the protected domain and the unprotected outside. Firewalls can screen traffic flowing both into and out of the protected domain, rejecting any data that do not conform to the policy the firewalls are configured to implement. They can protect against many kinds of attacks, including the following:

- Unauthorized internal access: An unauthorized user outside the protected domain seeks access to data or services inside the domain (for example, protecting company confidential information from outsiders).

- Compromising authentication: An external user obtains access privileges of an insider by fooling the authentication mechanism (for example, copying and reusing a password).
- Unauthorized external access: An internal user seeks access to unauthorized data or services outside the domain (for example, employees accessing leisure-time sites while on company business).
- Spoofing: The source address of data falsely makes it appear to come from a trustworthy source (for example, purporting to be a host within the protected domain).
- Session stealing: A session for one purpose is subverted to another purpose (for example, when an e-mail connection in progress is converted to one that opens a file transfer operation).
- Tunneling: An attack uses a subverted host to conceal the attack.
- Flooding: The attacker tries to overload an internal host with requests from the outside.

In addition, firewalls can maintain audit logs that can indicate after the fact what kind of damage occurred in case the firewall did not block an attack.

Firewalls implement either a simple blocking technology, proxy, or real service. In blocking, the firewall inspects incoming traffic and decides which data items to admit, perhaps admitting e-mail but blocking remote logins. A proxy is a limited-functionality emulator for a service that protects an internal host providing the actual service. For example, a firewall could provide a proxy FTP service, reflecting certain commands to the inside for actual implementation by the protected hosts, and blocking others (which might do things such as try to begin access to a protected directory). A firewall may instead provide a full service itself, such as reporting the status of only those known users that the protected organization wants to have reported perhaps shielding (because the firewall is not configured to know) the names or addresses of some internal users.

A more complex firewall structure uses two or more separated firewalls to implement a virtual private network. Suppose an organization has two or more sites connected by an insecure medium such as the Internet. If the organization uses encryption implemented by the two firewalls for all traffic between them, the encryption can effectively provide a private network, protected against attacks that target confidentiality and integrity:

- Steven Lipner, Technical Director, Technical Resource Center, Mitretek Systems
- Teresa Lunt, Program Manager, US Defense Advanced Research Projects Agency
- Ruth Nelson, President, Information System Security
- William Neugent, Chief Engineer, Infosec Division, The Mitre Corp.
- Hilarie Orman, Program Manager, US Defense Advanced Research Projects Agency
- Marcus Ranum, Chief Executive Officer, Network Flight Recorder, Inc.
- Roger Schell, Senior Development Manager, Network Security, Novell
- Eugene Spafford, Professor and Director of Coast Laboratory, Department of Computer Sciences, Purdue University

For an overview of firewalls and their major components, refer to the boxed text above.

DC/CP: Who should use a firewall?

Nougent: In most cases, organizations needing a firewall are those that have information or systems that they wish to protect from the Internet or from any network to which they are connected. If you do not mind your data or software being stolen, or your system modified or destroyed, or your company being publicly embarrassed, then you do not need a firewall. Our experience with customers is that the threat of public embarrassment is a particularly good motivator. Some companies use firewalls to control what their employees do on the Internet, to avoid the employees using services that undermine corporate security, or even to avoid employees leaving company "footprints" in awkward places.

Ranum: A firewall is appropriate for anyone connecting two networks that should be separate for security or administrative purposes. Examples are people connecting production and R&D. networks, people connecting networks to the Internet, and people connecting their networks to their business partners' networks.

DC/CP: How effectively do firewalls protect against Internet intrusions? That is, what do they protect against and what do they not protect against?

Lipner: A properly configured firewall can pretty well ensure that outsiders on the Internet cannot break into the protected network and access restricted services and data.

Lunt: General wisdom is that these firewall products are not very effective, but I have not seen any systematic studies that have tried to prove this.

Nougent: Look, how effective are exercise bikes? The point is, it depends on how you use them. And once you start, you have to keep working at it to retain the benefits. Firewalls change your way of life; they might also take some of the fun out of life by disallowing some risky services. The Puritans would have loved firewalls.

Haigh: Firewalls control the flow of services between a site and the Internet outside, and the direction in which these services flow. A firewall's protection weakens as the threat becomes more deeply embedded within the data it carries. For instance, almost any flavor of static or dynamic packet filtering can protect against basic network attacks. By limiting outside access to secured servers or servers sitting in a demilitarized zone outside the enclave, most firewalls can also reduce the risk of outside clients attacking vulnerable servers within the enclave. By integrating additional features, firewalls can provide additional forms of protection. For instance, a firewall that supports strong user authentication can enforce additional restrictions on Internet access, controlling which users have access and what types of activities users may perform.

Spafford: Firewalls are needed because most host software is poorly protected and badly designed. Basically, firewalls can be effective at helping to protect a site, but they are not foolproof, nor are they equally effective against all threats. I often compare firewalls to airbags. They help protect you against head-on collisions, but you must also be responsible, put on your seat belt, and not drive into a wall at 80 miles per hour. It does not help if you get hit from the side, if someone sets your seat on fire, or if your car has an airbag but no brakes.

DC/CP: Do firewalls have any benefits beyond the protection of internal networks from outside intrusion?

Gong: One added benefit is centralized administration: corporate policy can be configured and managed at a single point.

Lunt: The problem, though, with using a firewall as a single point of security management is that many organizations do not know all the points of access between their internal networks and external networks. So we should not expect a firewall to protect against all such access.

Orman: Firewalls offer the attractive advantage of network security administration that scales well as an organization grows, but they also reduce an organization's ability to provide network services to others.

Neugent: Many years ago, before we had a firewall, foreign hackers used our corporate network to attack our customers and others. At that time, my company had access to the global network and many of our customers did not. We permitted unrestricted access to our corporate networks so our customers could use our global connections. There were no access controls. In theory, customer usage of our corporate network was light and inexpensive for the good PR we got in return, or so we thought. The foreign hackers totally changed our thinking. Since then, we have not only installed a corporate firewall, we have also professionalized our information systems security (infosec) overall. We are still not impregnable because we interact with the Internet, and that interaction entails some unavoidable risk. Of course, some of our work is too sensitive to put at such risk, so we do that work on physically isolated systems.

Anderson: If one assumes an Internet environment where the "threat" is high school students interested in "exploring", then maybe (a big maybe) a firewall would effectively keep them out. If the threat is moderately serious, one where the adversary can monitor traffic to or from a target system in real time, then firewalls do little to protect you against anything. Even here, however, the firewall may bear the brunt of an attack and be able to prevent loss of important data within the enclave.

DC/CP: Oh, so a firewall can serve as a first line of defense?

Nelson: Yes, but the filtering needed to implement this protection is difficult to achieve.

Schell: The only information that the filtering program (the essence of a firewall) has to work with is the datagram in question and the rules setup in advance by the administrator. This lets the administrator block all "direct" access (that is, access from outside the enclave) or block all traffic believed to be "bad". However, because neither the TCP- nor IP-level protocols provide for authentication of header fields, the effect of this filtering is inherently limited: only stereotypical intrusions can be intercepted. In particular, the basic mechanism cannot enforce any policy depending on a strongly authenticated user, host, or destination. Moreover, the filters themselves require constant maintenance as new flaws in the protected host platforms or services are uncovered. To be effective, this intelligent filter must emulate the application or service in such detail that, for all practical purposes, it becomes the service.

Brand: Attacks such as spoofing, tunneling, and session stealing are preventable if the organization behind the firewall is willing to "bite the bullet", take a restrictive approach, and make use of features such as encryption and strong authentication. For instance, Trojan horse attacks often arrive embedded in attachments to e-mail or java applets. Some firewalls have the capability to disallow e-mail attachments and some can be configured to block importation of applets into the protected net. Though preventable, data-driven attacks still get through.

Gong: Actually, firewalls are ill-suited to handle executable content. Content filtering can only go so far. Executable content can arrive at a client machine via many means, and a firewall cannot discover and filter all of these. For example, executable code can be encapsulated inside an e-mail message, even without attachments!

Anderson: Firewalls do not prevent external attacks on a network. They do not prevent denial of service to any host or to the internal network as a whole. They do not prevent unauthorized transmission of sensitive data to the outside. (They might limit transmissions to known addresses, but that assumes no one else is listening. Not a very realistic assumption, unfortunately.)

DC/CP: But is not the purpose of a firewall to block all malicious traffic?

Neugent: The role of a firewall is not to plug the leaks in the dike, but to keep out most of the water while you create the leaks you want.

Lipner: It is possible to open a hole in a firewall to a service such as a network file server (NFS) that can unintentionally expose the inside net or allow operations such as downloading hostile Java or ActiveX. The solution to this problem would be to block NFS or run it over a virtual private network (VPN). You could also block Java and ActiveX at the firewall or restrict access to signed applets from trusted sources.

Neugent: One neat thing about firewalls is that they enable you, within your intranet, to take risks that you are not willing to take on the open Internet. You can use the latest nonsecure products and protocols. You still need some security within the intranet, of course, but basically the firewall enables you to do things "at home" that you would not do in public. Firewalls reduce risk by allowing a small set of protocols through the firewall and blocking all others.

DC/CP: What are the limitations of firewalls?

Ranum: Firewalls rely on the end user to configure them correctly. Many sites with firewalls have been broken into not because of flaws in the firewall, but because whoever set it up let too much through. So, while a correctly configured firewall will protect you against most attacks from the Internet, an incorrectly configured firewall may offer no protection at all.

Scholl: Firewalls cannot possibly prevent attacks they do not know about, which means they will always be Band-Aids.

Lipner: The list of things that a firewall does not protect against constitutes a work program for system and security managers: Firewalls do not protect against malicious insiders; they do not protect against connections to the Internet that do not run through the firewall—for example when a user establishes a dial-up Internet service provider connection to his or her desktop—and they only protect services and operations if they are configured to do so.

Brand: The malicious actions of an employee are unlikely to even get logged by the firewall. Even fax lines to a user's workstation can be used by an attacker to circumvent a firewall. In addition, firewalls do not provide nonrepudiation.

Neugent: Firewalls do not protect against some data-driven attacks or against newly discovered vulnerabilities of allowed protocols. Mobile code embedded in objects can be difficult, if not impossible, to filter or block. Session stealing is difficult to prevent unless the firewall includes some form of cryptography-based VPN technology. In short, firewalls do a good job of defending against the many Internet "ankle biters", but given the services typically allowed by firewalls, they are certainly not a sufficient defense against a technically savvy adversary who is determined to break through. (So a natural complement to a firewall is an organizational policy in which you try to avoid ticking off technically savvy people.)

Gong: As a practical concern, firewalls create the need for proxies that complicate application software. They are also expensive to install and maintain, and may not be suitable for small companies or branch offices.

DC/CP: Do users have an unwarranted sense of security in using firewalls?

Anderson: The fact is, firewalls are best at limiting the normal domain of discourse to a designated set. They are a control that is effective only against passive or benign environments. If I am able to observe the two-way traffic to or from a target system, I can do pretty much anything to the firewall and the applications supported on the target host/network. At the very least, I could disrupt the traffic it carries and blame it on a presumed known “friend”.

Oman: The Internet world is not a static one, and current trends are toward application-dominated environments with heavy reliance on networking. This moves the structure, location, and arrangement of security countermeasures into a realm that is difficult for firewalls to cope with, a disadvantage that firewalls probably cannot overcome.

Spafford: Unfortunately, firewalls do not protect against sloppy coding by firewall vendors, or misconfiguration or poor policy decisions by administrators.

Schell: That is because firewalls themselves are computers and are therefore subject to attacks that prey on a weak implementation. Their implementations most often lack a security perimeter, and even if they do possess one, it is not possible to determine whether their policy enforcement is performed correctly. This leaves open the very real possibility of malicious or flawed software being a significant part of the firewall implementation itself.

DC/CP: Would you recommend firewalls for connecting sensitive domains? For example, if Ford and General Motors were to form a joint venture, would you advise them to use one (or more) of today’s firewalls to join their corporate networks, with all of their marketing, product, and financial information?

Ranum: No way. Sensitive data are not worth risking for the convenience of firewall or network connectivity. A lot of people with seriously important data are at risk because of the illusion of security and the convenience of a firewall. It is insane to risk corporate jewels just so staff can surf playboy.com.

Spafford: I advise people not to even think about hooking up all their corporate resources. Putting everything together on a network makes all that information potentially available to everyone else. In your example, I would certainly recommend a firewall. But I would also recommend setting up separate networks so that not everything is exposed through the firewall.

Lunt: But some of this is already being done today, where a large corporation shares a network infrastructure with customers, subcontractors, and others. They see this as an unavoidable aspect of doing business.

Scholl: You should consider firewalls only if you have medium-value data to protect. Presumably, if the value of your data are low, you would not want to pay the price to purchase and maintain a firewall. On the other hand, if your data have high value, then the risk of loss is too great to entrust to a firewall. In other words, you should not entrust to a firewall the protection of data that are subject to attack by a determined penetrator. And it would be a mistake to believe that a firewall architecture can eventually be "perfected" to approach such a solution.

Haigh: Just as every large enterprise has some doors locked against otherwise trusted employees, every large network needs internal security to ht-nit unnecessary access to valuable resources. Studies have shown that "insiders" are the largest source of security problems. With the advent of extranets, the definition of an insider is also changing significantly. Business partners are insiders whose access should be severely limited. Thus it is necessary to build stronger security inside the enclave.

There are really three dimensions to consider when connecting to the Internet. What is the value of the efficiencies that can be gained and/or the new markets that can be tapped as a result of connection. What would be the cost of recovery from a breach of security if there were no protection or only limited protection? And what would be the cost of a security system that would reduce the likelihood of a security breach to an acceptable level? The security system will almost certainly include firewalls as well as other technical security mechanisms and modifications to business processes.

DC/CP: So a firewall is just part of a comprehensive security approach. What else is needed?

Spafford: Firewalls do not contribute to security education and awareness, nor do they help with physical security and employee screening.

Neugent: This question reminds me of the old saying, "To someone with a hammer, every problem looks like a nail". That is, it starts with the solution and asks where to stick it. What everyone on a network does need is an implemented security policy. A firewall is just one option as a component of such a security policy. Cost, security objectives, service objectives, and system architecture combine to determine if a firewall is needed. Fundamentally, if the security policy you want to implement is weak, then the world's strongest firewall will not help you (unless what you really seek is false reassurance, in which case it is terrific).

Lipner: Firewalls are an alternative to "security in depth", in which an organization ensures the security of each individual system on its inside network. Security in depth is almost always impractical because of the number and variety of systems on the inside network, the constant change of system configurations, and the varying competence and commitment of the system managers responsible for systems on the inside net.

DC/CP: Are firewalls an adequate alternative to security in depth?

Gong: By definition, a firewall cannot understand or deal with the security needs of a full-featured operating system. Otherwise, the firewall itself would be a secure operating system and one would not need to run the latter.

Ranum: A firewall and security in depth is the best alternative. A firewall that is installed in place of thinking about or planning security is no better than none at all. Most of the firewalls being installed today are the knee-jerk reactions of people trying to protect themselves.

Lunt: Many places are installing firewalls to show they have taken all reasonable precautions, that is, as liability protection for insurance purposes.

DC/CP: What advice would you give to someone on selecting or using a firewall product? Are there better alternatives?

Lipner: Do not start by focusing on the firewall. Look at your security policy (if you do not have one, write one). Look at the services your users will need to do their jobs, and the threats to your network and those services. Do you need to surf the Web? Do users need to log in (telnet) to inside systems or just check mail? Do you need to extend "inside" status to systems or networks across the Internet? All of those needs imply threats to be countered, security policies and standards for firewall configuration and operation, and features that will need to be present in the firewall you select.

Brand: And make your network security policy as restrictive as the organization will accept. Do this by getting input from end users and department managers as to what services they want and what they need. Ask the system administrators what services are actually being used. Document everything! Create a complete list of what is needed today and wanted for tomorrow. Gather information on the security issues for each item on the list. Then, service by service, determine if the security issues outweigh the need for the service. The result of this process will be a well-designed and defensible security policy. Have the appropriate management sign off on it. (Get as many signatories to the policy as possible. This can help when someone screams about no longer having access to their favorite Web site.)

Nebson: Firewalls can be most effective if the network behind them has a clear, restrictive security policy, of a form compatible with the capabilities of a firewall. Exceptions cause problems, as does flexible connectivity. One solution that helps is to use several layers of firewall protection, with Web servers and other heavily connected computers in an intermediate ring. Servers must still have their own protection. Unfortunately, the hardware can be expensive, and performance suffers.

Brand: Dozens of vendors have products marketed as firewalls. It can be difficult to weed though all of them. Compare the network security policy to the product reviews. Match the policy to the advertised features.

Neugent: One common problem we encounter is organizations using so many intrinsical insecure services over the network that one cannot install a firewall without seriously impeding their operation. Actually, there might be hope even for these organizations. Firewalls are mainly intended to prevent attacks. It is becoming possible to use intrusion detection to detect attacks, and maybe only deactivate or restrict those insecure services (using a firewall) when adversaries attempt to exploit them. Substantial improvements are needed in intrusion detection technology before this approach becomes reasonably effective, however. Intrusion detection can be hosted on your firewall or behind it, where it gives you a second line of defense against outsiders as well as against malicious internal attacks.

Anderson: Understand what you are using the firewall for. Do not overload it with "features" that are meant to overcome the lack of security within the network ostensibly protected by the firewall. Even though the machines inside the firewall have security features, they lack the means to protect themselves from malicious attacks. Consequently, when these attacks get past the firewall, there is nothing the firewall can do to help.

Neugent: My advice is to pretend you are Napoleon, only taller. Think big. Do not just solve your firewall problem. Think through your overall infosec needs before choosing such a crucial element as a firewall. And think through your communications needs as well. Do not just entrust this problem to the security geeks. The firewall has to fit in with the rest of the network architecture. In fact, a company's Rut "firewall" is often more than just a box. It also includes hardened hosts (for necessary but vulnerable services such as mail), inspection stations (to provide some protection against viruses), and other components.

Speaking of Napoleon, one of the critical question in fielding a firewall is, "Who's in charge?," We have a customer who has evolved an architecture with four firewalls from four different vendors, all with different policies and no way to manage or even monitor them centrally. The individual offices thought they were independent fiefdoms but then realized that, since they are all part of one organization, they are all interdependent.

Sometimes the hardest (and most valuable) aspect of fielding a firewall is figuring out what is "inside" and what is "outside" and who is responsible for guarding the inside.

True, these are not glitzy technical considerations just dumb old management issues. Yet this is fortunate, because it gives managers something to do while techies work out the more entertaining parts of the problem. More important than what a given firewall does today is how well it can be expected to improve and keep up with fast moving developments. So look at the vendor, not just at the box. It is a relationship, not an isolated purchase.

Brand: Request evaluation copies of the most promising firewalls and, if possible, do your own evaluations. Many firewall vendors offer 30-day evaluation licenses. Create a small LAN testbed made of a few PCs or workstations and hubs and try each out for a few days. Look for ease of installation, flexibility in configuration, usefulness of audit, well-written documentation, and availability of customer support.

Ranum: The products on the market today are all basically the same. There is a lot of hype surrounding the details, but any of the well-known firewall products is sufficient to protect a network, if the firewall is configured right. So, if someone is using a firewall product, the most important choice they can make is to learn what they are doing when they set it up. Making a reasonable selection of the services that are available is difficult and requires knowledge and wisdom.

DC/CP: At some point, potentially, every organization will have one or more firewalls; what then? What security problems should the community be looking at next?

Anderson: The skeptic in me says the same problems that existed before firewalls.

Spafford: I agree with Jim. We do not understand how to build policy, maintain integrity, effectively audit system use, detect misuse, or deal with a host of other problems.

Lipner: Some services are hard to support securely through a firewall, such as NFS file services or ActiveX or Java applets. Right now, for these kinds of services I would say either block them entirely or run them on a VPN implemented via encryption.

In the long term, one important trend will be the evolution of protocols that are "firewall friendly", so that the firewall can exercise its controls, and the combination of firewall and inside systems can operate securely. This is an especially challenging area, since encrypted protocols mean that firewalls cannot read the contents and exercise fine-grained control.

Neugent: I agree. New protocols and applications such as collaborative computing and mobile code sometimes use fundamentally insecure approaches. Advances such as firewall-based verification of digitally signed objects will help mitigate risk.

Brand: And we need research to develop content filtering to recognize and stop the importation of subversive code.

Spafford: Firewalls are not going to fully address the problem of downloading dangerous content. That will remain an area of concern.

Orman: The visible result of this is the continuing proliferation of viruses and the exploitation of major security holes in popular applications. Firewalls are ineffective against these problems, and it is not evident that they can ever keep pace with the tide of networked applications.

Gong: Firewalls are only one piece of the puzzle in solving the whole spectrum of Internet security problems. They will never be a substitute for secure operating systems.

Nelson: Traditional approaches to computer security also do not protect users against bad application code; they just keep data from "leaking" as a result of its execution. It is critical to address this functional security problem, to provide users the assurance that their application software will do what it is expected to do.

Lunt: The community should still be looking at how to connect to the Net safely, even if everyone has firewalls. We need technologies to allow us to create dynamic VPNs that can, with assurance, allow us to enforce a flexible set of policies on top of the policies of the individual enclaves in the association.

Haigh: Firewalls apply protections at the perimeter, and that is never enough. Configuration management and intrusion detection are the secondary issues once basic perimeter protection is in place. The complexity of configuring and managing security is driven by the wide spectrum of protocols and potential attacks. To stay ahead of the attackers, the firewall administrator must be familiar with the latest attacks and how to configure the firewall to defend against them.

Spafford: Two more problems will become more critical in the next few years: how to investigate computer misuse to apprehend malfeasors, and how to protect against a wide variety of denial-of-service attacks.

Lipner: Another area that will become important is providing better tools for securing internal networks. After all, as we said, most security breaches still result from the actions of authorized insiders. I dismissed securing the inside net as "security in depth", and I do not think internal security will ever be as strong as firewall security. But a combination of central management, central monitoring, and distributed security control components on individual systems can surely help.

Neugent: We had a customer recently, who, on detecting an attack, disconnected the organization's link to the Internet through the firewall. The customer was quite surprised when hardly any users complained. It turned out that many users had their own connections to the Internet. So we need better tools for mapping internal networks and looking for back doors around firewalls. Increased use of public-key infrastructures and application-level encryption will improve security in some ways, but it will also make it harder for firewall managers to know what users are really doing.

Orman: It is unlikely that every organization will have a firewall. Many organizations have noncentralized security management and are capable of managing an open system and protecting individual system. We should look to a future in which every machine is its own firewall.

Schell: The bottom line is, there is no magic pill for security. Solutions will come from building systems (including firewalls) that are trustworthy and applying reference monitor principles to enforce a precisely defined security perimeter with a systematically inspectable implementation.

19

A New Model for Probabilistic Information Retrieval on the Web

GABRIEL MATEESCU

Research Computing Support Group,
National Research Council Canada,
Ottawa, Ontario, Canada

MASHA SOSONKINA

Scalable Computing Laboratory,
Ames Laboratory, Ames, IA, USA

PAUL THOMPSON

Thayer School of Engineering,
Dartmouth College, Hanover, NH, USA

ABSTRACT*

Academic research in information retrieval did not make its way into commercial retrieval products until the last 15 years. Early web search engines also made little use of information retrieval research, in part because of significant differences in the retrieval environment on the Web, such as higher transaction volume and much shorter queries. Recently, however, academic research has taken root in search engines. This paper describes recent developments with a probabilistic retrieval model originating prior to the Web, but with features which could lead to effective retrieval on the Web. Just as graph structure algorithms make use of the graph structure of hyperlinking on the Web, which can be considered a form of relevance judgments, the model of this paper suggests how relevance judgments of web searchers, not just web authors, can be taken into account in ranking. This paper also shows how the combination of expert opinion probabilistic information retrieval model can be made computationally efficient through a new derivation of the mean and standard deviation of the model's main probability distribution.

1. INTRODUCTION

Academic research in information retrieval did not make its way into commercial retrieval until the last 15 years when products such as Personal Librarian (Koll, 1981) or, later, the ranked retrieval mode of West-law, WIN (West Publishing Company, 1993) became available. Early web search engines also made little use of information retrieval research, in part because of significant differences in the retrieval environment on the Web. Two main differences from earlier retrieval paradigms include higher transaction volume and much shorter

*Support for Paul Thompson for this research was provided by a Department of Defense Critical Infrastructure Protection Fellowship grant with the Air Force Office of Scientific Research, F49620-01-1-0272.

queries. More recently, however, academic research has taken root in search engines such as Google (Brin et al., 1998).

This paper describes recent developments with a probabilistic retrieval model that originated prior to the Web, but which has features that may lead to effective retrieval on the Web. Just as graph structure algorithms make use of the graph structure of hyperlinking on the Web (see, for example, Brin et al., 1998, Kleinberg, 1999), which can be considered a form of relevance judgments, the model of this paper shows how the relevance judgments of web searchers, not just web authors, can be taken into account in ranking. This paper also shows how the Combination of Expert Opinion probabilistic information retrieval model can be made computationally efficient through a new derivation of the mean and standard deviation of the model's main probability distribution.

2. BACKGROUND

The Bayesian Combination of Expert Opinion (CEO) approach to probabilistic information retrieval was first described by Thompson (1986, 1990a,b). The CEO model is a generalization of the unified probabilistic retrieval model developed by Robertson, Maron and Cooper (1982), also known as the RMC model. The unified model, called Model 3 in Robertson et al. (1982), was an attempt to combine the models of probabilistic information retrieval developed by Maron and Kuhns (1960), referred to as Model 1, with the probabilistic retrieval model developed by Robertson and Sparck Jones (1976), van Rijsbergen (1979), Croft and Harper (1979), and others, referred to as Model 2. As it has been the case with most probabilistic
retrieval models, these models were based on the use of point probabilities, rather than on probability distributions.

The CEO model, by contrast, provides a probability distribution for a document's being judged relevant by a particular user. Both the mean and standard deviation of the distributions are needed in the CEO model for the combination process, as well as the ranking of retrieved documents by probability of relevance. In early accounts of the model

(Thompson, 1990a,b), it was not shown how the mean and standard deviation, or variance, of these distributions could be computationally implemented. This paper shows how the mean and standard deviation of the CEO model's distribution can be computed and how the CEO model can be applied to Web document retrieval.

The unified probabilistic retrieval model, Model 3, was developed so that probabilistic evidence of relevance from the two earlier probabilistic models, Models 1 and 2, could be combined in order to produce a more accurate ranking of documents. As stated in the probability ranking principle (van Rijsbergen, 1979):

> If a reference retrieval system's response to each request is a ranking of the documents in the collection in order of decreasing probability of relevance to the user who submitted the request, where the probabilities are estimated as accurately as possible on the basis of whatever data have been made available to the system for this purpose, the overall effectiveness of the system to its user will be the best that is obtainable on the basis of those data.

There were several unresolved issues with the RMC version of Model 3. Robertson (1984) has shown that the term independence assumptions on which the model is based lead to inconsistencies. Moreover, the RMC version of Model 3 did not support relevance feedback. The CEO model, which is based on Model 3, was developed to overcome these difficulties, as well as to provide a general probabilistic retrieval model that could combine probabilities from multiple probabilistic retrieval models, not only the two models unified by the RMC model. In particular, it explored the use of subjective probabilities provided by indexers or searchers (Thompson, 1988).

3. RELATED RESEARCH

The past decade has seen much research on the combination of results from multiple retrieval algorithms, representations of text and query, and retrieval systems (Croft, 2000). The motivation for this research has been provided by several empirical studies showing that different algorithms, representations, and systems provide substantially different, though overlapping, sets of relevant documents (Croft and Harper, 1979; Katzer et al., 1982; McGill et al., 1979; Saracevic and Kantor, 1988). This activity has manifested itself both in academic research and in the commercial development of various Web metasearch engines (Aslam and Montague, 2001; Belkin et al., 1995; Manmatha et al., 2001; Selberg and Etzioni, 1999; Selberg, 1999).

Combination of models has also been an active area of research in other fields, including statistics (Hoeting et al., 1999; Moerland, 1999, 2000), statistical decision theory (Clemen and Winkler, 1999; Lindley, 1983; Roback and Givens, 2001), statistical pattern recognition (Jain et al., 2000; Xu et al., 1992), machine learning (Freund and Schapire, 1997; Littlestone and Warmuth, 1992; Lewis et al., 1996; Schapire and Singer, 1998) and neural networks (Hashem et al., 1994, 1997; Hofmann and Puzhica, 1998; Hofmann et al., 1999; Jordan and Jacobs, 1994; Jacobs et al., 1991; Tumer and Ghosh, 1996, 1999).

Many researchers have applied machine learning techniques to automatic text categorization or clustering, see, for example, Lewis et al. (1996). Mathematical techniques new to document retrieval, such as singular value decomposition, or latent semantic indexing, have also been applied (Deerwester, 1990). More recently, probabilistic variants of latent semantic indexing have been implemented as well (Hofmann, 1999, 2001; Papadimitriou et al., 1998).

4. THE COMBINATION OF EXPERT OPINION MODEL

The CEO model is a version of Model 3 that uses probability distributions, while the RMC version uses point probabilities. Furthermore, unlike the RMC model, which is based on point reconciliation, the CEO model applies Lindley's approach (Lindley, 1983) to reconciliation of probability distributions to probabilistic information retrieval. In this Bayesian model, a decision maker with an initial, or prior, probability (or distribution) for some event or parameter θ, consults n experts who provide their probabilities (or distributions) as evidence with which to update the decision maker's prior probability distribution via Bayes' theorem to obtain a revised, or posterior, probability (or distribution). In the CEO approach, there are two levels of combination. At the upper level, the probabilistic information retrieval system is considered the decision maker, and Models 1 and 2 the experts. At the lower level, Models 1 and 2 themselves are derived from CEO. The indexer, or user, in Model 1, or 2, respectively, is seen as a multiple expert—an expert with respect to each use or document property. Each expert, or decision maker, is estimating θ^*, the chance of relevance of a document with respect to a query, i.e., the long-run relative frequency of success in a Bernoulli sequence of relevance judgments of users for documents. Each Bernoulli sequence is different, but there is a common subsequence that underlies each, so that each expert, or decision maker, can be seen as estimating θ^*, for the underlying subsequence. The parameter actually used in the model is θ, the log-odds function of θ^*, i.e., $\theta = \log[\theta^*/(1-\theta^*)]$.

In the CEO model, the evidence provided to the decision maker (or information system) by the models being combined is the set of mean values and standard deviations of the distributions provided by the experts consulted by the decision maker. Let $p(m|s, \theta)$ denote the decision maker's probability distribution for the expert saying that the mean for the log-odds of the chance of relevance is m, given that the expert provides the standard deviation s for the log-odds of the chance of relevance and given the true value of θ. The decision maker's opinion of the experts' expertise, i.e., the weighting of the experts' evidence, is expressed by assuming that $p(m|s,\theta)$ is normal with mean $\alpha + \beta\theta$ and standard deviation $\gamma\ s$, where α, β and γ are parameters that can be determined either through data (relevance judgments) or a decision maker's subjective belief ($\gamma\ s$ is the result of modifying s by the factor γ, as called for in Lindley's model for reconciliation (Lindley et al., 1983), on which the CEO model is based).

A simplified version of the CEO model was used in the first and second Text Retrieval Conferences (TREC) (Thompson, 1993, 1994). In this version, the mean and standard deviation of the model's main distribution were calculated using approximate techniques. More importantly, relevance feedback was not incorporated in TREC 1. In TREC 2, a form of relevance feedback was used. The ranked retrieval models combined in the TREC 1 system were weighted by their performance. Unfortunately, due to the many changes made to the models between TREC 1 and TREC 2, the models' performance on TREC 2 was not well predicted by their Model 1 performance.

5. RELEVANCE FEEDBACK

Relevance feedback, i.e., the incorporation of users' judgments as to the relevance of retrieved documents to their information needs, presented a problem with pre-web retrieval. Laboratory experiments showed that large gains in performance, in terms of precision and recall, could be gained through use of relevance feedback (Ide and Salton, 1971). On the other hand, it was assumed that it would not be possible to induce users to provide relevance judgments. Westlaw's WIN was introduced without a relevance feedback capability (West Publishing, Company, 1993). By contrast, Lexis–Nexis' Freestyle and some web search engines introduced commands that provided relevance feedback based on a single document, rather than a set of relevant documents. These are often called "more like this" commands, where a user selects a single highly relevant returned document and the system returns similar documents. In the TREC conferences and other experimental settings, use has been made of pseudo-relevance feedback, where the top n documents are assumed to be relevant and relevance feedback is calculated as though these n documents had actually been judged relevant (Sakai et al., 2001). As pointed out by Croft et al. (2001), early work on relevance feedback was done with collections of abstracts and results with full text documents have not been as good as was anticipated.

In addition to this type of more or less traditional relevance feedback, new forms of relevance feedback have emerged, including implicit relevance feedback, e.g., systems such as Direct Hit (DH) (which provides relevance feedback based on mining a user's clickstream), recommender systems (Herlocker, 2001), and rating systems (Dellarocas, 2001).

Relevance feedback is usually seen as taking place during a single user's search, but relevance feedback has also been considered in more persistent ways, e.g., in dynamic document spaces (Brauen, 1971). In dynamic document spaces, a user's relevance judgments permanently modify the weights of index terms associated with documents.

6. PROBABILITY DISTRIBUTIONS IN THE COMBINATION OF EXPERT OPINION

As mentioned above, each probabilistic model, e.g., the indexer or the user, is making an estimate of θ for a common underlying Bernoulli subsequence of the overall Bernoulli sequence of viewings of documents by users. Because each model is making these judgments based on the conditioning information available to it, that model's judgment for the sequence's distribution is exchangeable, i.e., the distribution is invariant under finite permutations of its indices (de Finetti, 1974). A natural distribution to use for a parameter that ranges from 0 to 1, e.g., the proportion of successes in a sequence of relevance judgments, in the beta distribution (Bunn, 1984). It can be very simply updated with each relevance judgment. Graphically, the beta distribution can take many shapes, and is thus capable of expressing a wide range of opinions. The CEO algorithm uses a transformation of the beta distribution, the distribution of $\log[x/(1-x)]$, where x is a random variable with a beta distribution. It is this distribution, referred to as the transformed beta distribution, from which the mean and standard deviation need to be extracted in order to perform the combination of expert opinion and to probabilistically rank retrieved documents.

7. COMPUTING THE MEAN AND STANDARD DEVIATION OF THE TRANSFORMED BETA DISTRIBUTION

Let y be a continuous random variable whose distribution function is the transformed beta distribution. In this section, we derive expressions for the mean value and standard deviation of y.

The moment generating function of y is the function ψ defined by (see Thompson, 1990b):

$$\psi(t) = \frac{\Gamma(p+t)\Gamma(q-t)}{\Gamma(p)\Gamma(q)}$$

where $p,q > 0$, $-\infty < x < \infty$ and $\Gamma(x)$ is the Gamma function defined by

$$\Gamma(x) = \int_0^\infty t^{x-1}\mathrm{e}^{-1}\mathrm{d}t$$

7.1. Computation of the Mean Value

The mean value μ of the random variable y is derived as follows

$$\begin{aligned}\mu &= \left.\frac{\mathrm{d}\psi(t)}{\mathrm{d}t}\right|_{t=0} = \left.\frac{\mathrm{d}}{\mathrm{dt}}\frac{\Gamma(p+t)\Gamma(q-t)}{\Gamma(p)\Gamma(q)}\right|_{t=0} \\ &= \left.\frac{\mathrm{d}}{\mathrm{dt}}\frac{\Gamma(p+t)}{\Gamma(p)}\right|_{t=0}\left.\frac{\Gamma(q-t)}{\Gamma(q)}\right|_{t=0} + \left.\frac{\mathrm{d}}{\mathrm{d}t}\frac{\Gamma(q-t)}{\Gamma(q)}\right|_{t=0}\left.\frac{\Gamma(p+t)}{\Gamma(p)}\right|_{t=0} \\ &= \left.\frac{\Gamma'(p+t)}{\Gamma(p)}\right|_{t=0} - \left.\frac{\Gamma'(q-t)}{\Gamma(q)}\right|_{t=0} \qquad (1)\end{aligned}$$

where we have used these facts

$$\frac{\mathrm{dt}(p+t)}{\mathrm{dt}} = 1 \quad \text{and} \quad \frac{\mathrm{dt}(q-t)}{\mathrm{dt}} = -1 \qquad (2)$$

Because $\Gamma'(x \pm t)|_{t=0} = \Gamma'(x)$ we get from (1)

$$\mu = \frac{\Gamma'(p)}{\Gamma(p)} - \frac{\Gamma'(q)}{\Gamma(q)} \qquad (3)$$

It can be shown (see Whittaker and Watson, 1990) that

$$\Gamma(x) = \lim_{n\to\infty} \frac{n^x n!}{x(x+1)\cdots(x+n)} \tag{4}$$

and it is easy to show that

$$\frac{n^x n!}{x(x+1)\cdots(x+n)} = e^{x(\ln n - 1 - (1/2) - \cdots - (1/n))} \frac{1}{x} \frac{e^{x/1}}{1+(x/1)} \frac{e^{x/2}}{1+(x/2)} \cdots \frac{e^{x/n}}{1+(x/n)}$$

Therefore, substituting in (4) we get

$$\Gamma(x) = e^{-Cx} \frac{1}{x} \prod_{n=1}^{\infty} \frac{e^{x/n}}{1+(x/n)} \tag{5}$$

where C is the Euler–Macheroni constant defined by the limit

$$C = \lim_{x\to\infty} \left(1 + \frac{1}{2} + \frac{1}{3} + \cdots + \frac{1}{n} - \ln n \right)$$

and the value of C computed with 10 decimal places is

$$C = 0.5772156649$$

Taking the logarithm of (5) and differentiating gives:

$$\frac{\Gamma'(x)}{\Gamma(x)} = -C - \frac{1}{x} + \sum_{i=1}^{\infty} \frac{x}{i(x+i)} \tag{6}$$

We are now ready to compute μ from (3):

$$\begin{aligned} \mu &= -C - \frac{1}{p} + \sum_{i=1}^{\infty} \frac{p}{i(p+i)} - \left(-C - \frac{1}{q} + \sum_{i=1}^{\infty} \frac{q}{i(q+i)} \right) \\ &= \frac{p-q}{pq} + \sum_{i=1}^{\infty} \frac{p}{i(p+i)} - \sum_{i=1}^{\infty} \frac{q}{i(q+i)} \end{aligned} \tag{7}$$

7.2. Computation of the Standard Deviation

The standard deviation σ^2 of y is defined as

$$\sigma^2 = E[y^2] - \mu^2$$

with $E[y^2]$ being the second moment of y that is computed using the formula:

$$E[y^2] = \left.\frac{d^2\psi(t)}{dt^2}\right|_{t=0} = \left.\frac{d^2}{dt^2}\frac{\Gamma(p+t)\Gamma(q-t)}{\Gamma(p)\Gamma(q)}\right|_{t=0}$$
$$= \left.\frac{\Gamma''(p+t)\Gamma(q-t) - 2\Gamma'(p+t)\Gamma'(q-t) + \Gamma(p+t)\Gamma''(q-t)}{\Gamma(p)\Gamma(q)}\right|_{t=0}$$

where we have used the facts (2). Thus

$$E[y^2] = \frac{\Gamma''(p)}{\Gamma(p)} + \frac{\Gamma''(q)}{\Gamma(q)} - 2\frac{\Gamma'(p)}{\Gamma(p)}\frac{\Gamma'(q)}{\Gamma(q)}$$

Formally differentiating (6), we get

$$\frac{\Gamma''(x)}{\Gamma(x)} = \left(\frac{\Gamma'(x)}{\Gamma(x)}\right)^2 + \frac{1}{x^2} + \sum_{i=1}^{\infty}\frac{1}{(x+i)^2}$$

then, substituting (6) in the above relation, we obtain:

$$\frac{\Gamma''(x)}{\Gamma(x)} = \left(-C - \frac{1}{x} + \sum_{i=1}^{\infty}\frac{x}{i(x+i)}\right)^2 + \frac{1}{x^2} + \sum_{i=1}^{\infty}\frac{1}{(x+i)^2} \qquad (8)$$

The second moment of y is

$$E[y^2] = \left(-C - \frac{1}{p} + \sum_{i=1}^{\infty}\frac{p}{i(p+i)}\right)^2 + \frac{1}{p^2} + \sum_{i=1}^{\infty}\frac{1}{(p+i)^2}$$
$$+ \left(-C - \frac{1}{q} + \sum_{i=1}^{\infty}\frac{q}{i(q+i)}\right)^2 + \frac{1}{q^2} + \sum_{i=1}^{\infty}\frac{1}{(q+i)^2}$$
$$- 2\left(-C - \frac{1}{p} + \sum_{i=1}^{\infty}\frac{p}{i(p+i)}\right)\left(-C - \frac{1}{q} + \sum_{i=1}^{\infty}\frac{q}{i(q+i)}\right) \qquad (9)$$

Therefore, the standard deviation and the mean value can be computed approximately by replacing $\sum_{i=1}^{\infty}$ with $\sum_{i=1}^{n}$ in relations (7) and (9), then taking n large enough to meet a given convergence criterion.

8. DISCUSSION

The CEO model provides a probabilistic framework for combining probabilistic retrieval models. The model can be used with subjective probabilities provided, either explicitly or implicitly, by users. It can be used both within the context of a single search and over time. Search on the Web is different in various ways from traditional online document retrieval. Two of the main differences, higher transaction volume and shorter queries, are differences that can be taken advantage of by the CEO model. First, high transaction volumes mean that there are more documents being seen by users from which relevance judgments can be collected. Second, because queries are so much shorter, on average less than three words per query, as compared to seven words more typical or traditional online retrieval, it is important to extend the focus of probabilistic models beyond words in documents and queries. As mentioned above, algorithms such as HITS (Kleinberg, 1999) or Page Rank (Brin et al., 1998) extend the focus to hyperlinking. The CEO model shows how this focus could be further extended to user's relevance judgments, whether explicit or implicit.

The statistical model of the reconciliation of probability distributions, on which the CEO algorithm is based, has seen significant development in recent years, e.g., (Roback and Givens, 2001). Related work has been done in machine learning, e.g., on the weighted majority voting algorithm and on boosting (Freund and Schapire, 1997; Littlestone and Warmuth, 1992; Lewis et al., 1996; Schapire and Singer, 1998), mixture models (Cohn and Hofmann, 2001; Hofmann and Puzhica, 1998; Hofmann, 1999, 2001; Hofmann et al., 1999; Jacobs et al., 1991; Jordan and Jacobs, 1994; Moerland, 1999, 2000; Manmatha et al., 2001), and Bayesian model averaging

(Hoeting et al., 1999). Text categorization and clustering have become significant application domains for machine learning research. Algorithms such as boosting (Schapire and Singer, 1998) and support vector machines (Joachim, 2001) have achieved good results with text categorization.

The focus of these new machine learning and related techniques has been on the document collection, not on the user and the user's information need. As noted by Papadimtriou et al. (1998), "The approach in this body of work (probabilistic information retrieval) is to formulate information retrieval as a problem of learning the concept of "relevance" that relates documents and queries. The corpus and its correlations play no central role. In contrast, our focus is on the probabilistic properties of the corpus." This focus on the collection ignores the probabilistic evidence provided by an analysis of the user and the user's information need. Relevance is better understood as a relation between the user's information need, which is represented by the query, and the intellectual content of the document, which is represented by the text of the document (Wilson, 1973) . While the text of queries and documents may model this latent, deeper structure, especially in the case of the document, user's relevance judgments (Croft et al., 2001) and mixed-initiative interaction (Haller et al., 1999) provide additional evidence of the user's information need. Much research in probabilistic information retrieval is currently focused on language models (Callan et al., 2001; Ponte and Croft, 1998; Ponte 1998). Language models are also mainly applied to collections, rather than users, though Lafferty and Zhai (2001) provide two language models, one for the document and one for the query, and perform retrieval by measuring the similarity of the two language models.

The CEO model predated much of the research discussed above in the fields of statistics, neural networks, and machine learning. Lindley's (1983) model of reconciliation of distributions, now called Supra-Bayesian pooling, on which the CEO model is based, is still one of the leading theories in the Bayesian approach to combining expert opinions (Roback and Givens, 2001). The basic framework of the CEO model appears to be sound, but the model still needs to be completely implemented and empirically tested. In the process of doing so it is likely that the model can be improved through the incorporation of some aspects of the more recent research discussed above. In particular, although there is long-standing precedence in the decision theory literature (Bunn, 1984) for using the beta distribution, as discussed above, to model expert opinion, it may be that techniques from Bayesian model averaging (Hoeting et al., 1999) could lead to more accurate modeling. With respect to representation of experts' opinion, the CEO model only requires a mean and standard deviation, not a specific distributional form. More generally, mixture models now being explored in the context of information retrieval, (Cohn and Hofmann, 2001; Hofmann, 1999, 2001; Manmatha et al., 2001), may foster new developments with the CEO model.

9. CONCLUSION

The probability ranking principle calls for taking all available evidence into account when probabilistically ranking documents in response to a user's request.

The CEO algorithm provides a formalism for taking all such evidence into account using Bayesian subjective decision theory. The theoretical strength of the CEO algorithm, its ability to easily incorporate relevance judgments and use the judgments to continuously tune its probability estimates, has also been its practical weakness. The success of recommender and similar systems in some domains, e.g., e-commerce, shows that implicit relevance judgments can be effective and may lead to settings where algorithms such as CEO, which rely heavily on relevance judgments, can be effective. Now that an efficient method of calculating the mean and standard deviation of the transformed beta distribution has been derived, the implementation of the CEO model will be facilitated.

REFERENCES

Aslam, J. A., Montague, M. (2001). Models for metasearch. In: Croft, W. B., Harper, D. J., Kraft, D., Zobel, J., eds. Proceedings of the 24th Annual Retrieval (SIGIR' 01). New Orleans, Sept. 9–13, ACM Press, pp. 276–284.

Belkin, N., Kantor, P., Fox, E., Shaw, J. (1995). Combining the evidence of multiple query representations for information retrieval. *Information Processing & Management* 31(3): 431–448.

Brauen, T. L. (1971). Document vector modification. Salton, G., ed. *Experiments in Automatic Document Processing*. Englewook Cliffs, NJ: Prentice-Hall, 456–484.

Brin, S., Motwani, R., Page, L., Winograd, T. (1998). What can you do with a web in your pocket? *Bulletin of the IEEE Computer Society Technical Committee on Data Engineering* 21(2):37–47.

Bunn, D. W (1984). *Applied Decision Analysis*. New York, NY: McGraw-Hill.

Callan, J., Croft, B., Lafferty, J. (2001). Proceedings of the Workshop on Language Modeling and Information Retrieval. Carnegie Mellon University, May 31–June 1. http://la.lti.cs.cmu.edu/callan/Workshops/lmir01/WorkshopProcs.

Clemen, R. T., Winkler, R. L. (1999). Combining probability distributions from experts in risk analysis. *Risk Analysis* 19: 187–203.

Cohn, D., Hofmann, T. (2001). The missing link: a probabilistic model of document content and hypertext connnectivity. *Advances in Neural Information Processing Systems (NIPS 13)*. Cambridge, MA: MIT Press 13:430–433.

Croft, W. B. (2000). Combining approaches to information retrieval. Croft, W. B., ed. *Advances in Information Retrieval: Recent Research from the Center for Intelligent Information Retrieval*. Chapter 1. Boston, MA: Kluwer Academic, 1–36.

Croft, W. B., Harper, D. J. (1979). Using probabilistic models of document retrieval without relevance information. *Journal of Documentation* 45(4):285–295.

Croft, W. B., Cronen-Townsend, S., Lavrenko, V. (2001). Relevance feedback and personalization: a language modeling perspective. In: Joint DELOS-NSF Workshop on Personalization and Recommender Systems in Digital Libraries. Dublin, Ireland, June 18–20, pp. 18–20.

Deerwester, S., Dumais, S. T., Furnas, G. W., Landauer, T. K., Harshman, R. (1990). Indexing by latent semantic analysis. *Journal of the American Society for Information Science* 41(6):381–407.

Dellarocas, C. (2001). Building trust on-line: the design of reliable reputation reporting mechanisms for online trading communities. Paper 101. Center for eBusiness@MIT.

de Finetti, B. (1974) (1974). *Theory of Probability: A Critical Introductory Treatment*. Vol. 1. New York: John Wiley.

Direct Hit. (2002). http://www.directhit.com.

Freund, Y., Schapire, R. E. (1997). A decision-theoretic generalization of on-line learning and an application to boosting. *Journal of Computer and System Science*. 55(1): 119–139.

Haller, S., McRoy, S., Kobsa, A. (1999). *Computational Models of Mixed-Initiative Interaction*. Boston, MA: Kluwer Academic.

Hashem, S. (1997). Algorithms for optimal linear combinations of neural networks. *International Conference on Neural Networks* Vol. 1, pp. 242–247.

Hashem, S., Schmeiser, B., Yih, Y. (1994). Optimal linear combinations of neural networks: an overview. IEEE World Congress on Computational Intelligence, IEEE International Conference on Neural Networks, Vol. 3, pp. 1507–1512.

Herlocker, J. (2001). ACM SIGIR Workshop on Recommender Systems Notes. http://www.cs.orst.edu/~herlock/rsw2001/workshop_notes.html.

Hoeting, J. A., Madigan, D., Raftery, A. E., Volinsky, C. T. (1999). Bayesian model avaraging: a tutorial. *Statistical Science* 14(4):382–417.

Hofmann, T. (1999). Probabilistic latent semantic indexing. In: Hearst, M., Gey, F., Tong, R., eds. Proceedings of the 22nd Annual International ACM SIGIR Conference on Research and Development in Information Retrieval (SIGIR' 99), Berkeley, CA, ACM Press, Aug. 15–19, pp. 50–57.

Hofmann, T. (2001). What people (don't) want. European Conference on Machine Learning (ECML).

Hofmann, T., Puzhica, J. (1998). Unsupervised learning from dyadic data. Technical Report TR-98-042, University of California, Berkeley.

Hofmann, T., Puzhica, J., Jordan, M. (1999). Learning from dyadic data. In: Kearns, M. S., Solla, S. A., Cohn, D., eds. *Advances in Neural Information Processing Systems*. Number 11. Cambridge, MA: MIT Press.

Ide, E., Salton, G. (1971). Interactive search strategies and dynamic file organization in information retrieval. In: Salton, G., ed. *Experiments in Automatic Document Processing*. Englewood Cliffs, NJ: Prentice-Hall, pp. 373–393.

Jacobs, R. A., Jordan, M. I., Nowlan, S., Hinton, G. E. Adaptive mixtures of local experts. *Neural Computation* 3:1–12.

Jain, A. K., Duin, R. P. W., Mao, J. (2000). Statistical pattern recognition: a review. *IEEE Transactions on Pattern Analysis and Machine Intelligence* 22(1):4–37.

Joachim, T. (2001). A statistical learning model of text categorization for support vector machines. In: Croft, W. B., Harper, D. J., Kraft, D., Zobel, J., eds. Proceedings of the 24th Annual International ACM SIGIR Conference on Research and Development in Information Retrieval (SIGIR' 01), New Orleans, Sept. 9–13, ACM Press, pp. 128–136.

Jordan, M. I., Jacobs, R. A. (1994). Hierarchical mixtures of experts and the EM algorithm. *Neural Computation* 6:181–214.

Katzer, J., McGill, M., Tessier, J., Frakes, W., Das-Gupta, P. (1982). A study of the overlap among document representations information technology. *Research and Development* 1(4):261–274.

Kleinberg, J. M. (1999). Authoritative sources in a hyperlinked environment. *Journal of the ACM* 46(5):604–622.

Koll, M. B. (1981). Information retrieval theory and design based on a model of the user's concept relations. Oddy, R. N., Robertson,

S. E., van Rijsbergen, C. J., Williams, P. W., eds. *Information Retrieval Research*. London: Butterworths.

Lafferty, J., Zhai, C. (2001). Document language models, query models, and risk minimization for document retrieval. In: Croft, W. B., Harper, D. J., Kraft, D., Zobel, J., eds. Proceedings of the 24th Annual International ACM SIGIR Conference on Research and Development in Information Retrieval (SIGIR' 01). New Orleans, Sept. 9–13, ACM Press, pp. 111–119.

Lewis, D. D., Schapire, R. E., Callan, J. P., Papka, R. (1996). Training algorithms for linear text classifiers. In: Frei, H. P., Harman, D., Schauble, P., Wilkinson, R., eds. Proceedings of the 19th Annual International, ACM SIGIR Conference on Research and Development in Information Retrieval (SIGIR' 96). Zurich, Switzerland, Aug. 18–22, ACM Press, pp. 298–306.

Lindley, D. V. (1983). Reconciliation of probability distributions. *Operations Research* 31(5):866–880.

Littlestone, N., Warmuth, M. (1992). The weighted majority voting algorithm. Technical Report UCSC-CRL-91-28, University of California, Santa Cruz.

Manmatha, R., Rath, T., Feng, F. (2001). Modeling score distributions for combining the outputs of search engines. In: Croft, W. B., Harper, D. J., Kraft, D., Zobel, J., eds. Proceedings of the 24th Annual International ACM SIGIR Conference on Research and Development in Information Retrieval (SIGIR' 01). New Orleans, Sept. 9–13, ACM Press, pp. 267–275.

Maron, M. E., Kuhns, J. L. (1960). On relevance, probabilistic indexing and information retrieval. *Journal of the ACM* 7(3):216–244.

McGill, M., Koll, M., Noreault, T. (1979). An evaluation of factors affecting document ranking by information retrieval systems. Final report for grant NSF-ISF-78-10454 to the National Science Foundation, Syracus University School of Information Studies.

Moerland, P. (1999). A comparison of mixture models for density estimation. Proceedings of the International Conference on Artificial Neural Networks (ICANN' 99).

Moerland, P. (2000). Mixtures of latent variable models for density estimation and classification. Research Report IDIAP-RR 00-25, IDIAP, Martigny, Switzerland.

Papadimitriou, C. H., Raghavan, P., Tamaki, H., Vempala, S. (1998). Latent semantic indexing: a probabilistic analysis. In: Mendelson, A., Paredaens, J., eds. Proceedings of the seventeenth ACM SIGACT-SIGMOD-SIGART Symposium on Principles of Database Systems (PODS' 98). Vol. 3, June 1–4, Seattle WA, ACM Press, pp. 159–168.

Ponte, J. (1998). A language modeling approach to information retrieval. PhD thesis, University of Massachusetts, Amhersts.

Ponte, J., Croft, W. B. (1998). A language modeling approach to information retrieval. In: Croft, W. B., Moffat, A., van Rijsbergen, C. J., Wilkinson, R., Zobel, J., eds. Proceedings of the 21st Annual International ACM SIGIR Conference on Research and Development in Information Retrieval (SIGIR' 98). Melbourne, Australia, Aug. 24–28, ACM Press, pp. 275–281.

Roback, P. J., Givens, G. H. (2001). Supra-bayesian pooling of priors linked by a deterministic simulation model. *Communications in Statistics-Simulation and Computation* 30:447–476.

Robertson, S. E. (1984). Consistency of the RMC model. Personal communication.

Robertson, S. E., Sparck Jones, K. (1976). Relevance weighting of search terms. *Journal of the American Society for Information Science* 27(3):129–146.

Robertson, S. E., Maron, M. E., Cooper, W. S. (1982). Probability of relevance: a unification of two competing models for document retrieval. *Information Technology: Research and Development* 1(1):1–21.

Sakai, T., Robertson, S. E., Walker, S. (2001). Flexible pseudo-relevance feedback for NTCIR-2. In: Proceedings of the Second NTCIR Workshop on Evaluation of Chinese and Japanese Text Retrieval and Text Summarization. Tokyo, Japan. Tokyo: National Institute of Informatics.

Saracevic, T., Kantor, P. (1988). A study of information seeking and retrieving. III. Searchers, searches, overlap. *Journal of the American Society for Information Science* 39(3):197–216.

Selberg, E. W. (1999). Towards comprehensive web search. PhD thesis, University of Washington.

Selberg, E., Etzioni, O. (1997). The MetaCrawler architecture for resource aggregation on the Web. *IEEE Expert* 12(1):8–14.

Schapire, R. E., Singer, Y. (1998). Improved boosting algorithms using confidence-rated predictions. Proceedings of the Eleventh Annual Conference on Computational Learning Theory, Madison, Wisconsin, July 24–26.

Thompson, P. (1986). Subjective probability, combination of expert opinion, and probabilistic approaches to information retrieval. PhD thesis, University of California, Berkeley.

Thompson, P. (1988). Subjective probability and information retrieval: a review of the psychological literature. *Journal of Documentation* 44(2):119–143.

Thompson, P. (1990a). A combination of expert opinion approach to probabilistic information retrieval, part 1: The conceptual model. *Information Processing & Management* 26(3):371–382.

Thompson, P. (1990b). A combination of expert opinion approach to probabilistic information retrieval, part 2: Mathematical treatment of CEO Model 3. *Information Processing & Management* 26(3):383–394.

Thompson, P. (1993). Description of the PRC CEO algorithm for TREC. In: Harman, D. K., ed. The First Text Retrieval Conference (TREC-1). volume NIST Special Publication 500-207. Washington, DC: US Government Printing Office, pp. 337–342.

Thompson, P. (1994) Description of the PRC CEO algorithm for TREC-2. In: Harman, D. K., ed. The Second Text Retrieval Conference (TREC-2). volume NIST Special Publication 500-215. Washington, DC: US Government Printing Office, pp. 271–274.

Tumer, K., Ghosh, J. (1996). Theoretical foundations of linear and order statistics combiners for neural pattern classifiers. Technical Report TR-95-02-98. The University of Texas at Austin.

Tumer, K., Ghosh, J. (1999). Linear and order statistics combiners for pattern classification. Sharkey, A., ed. *Combining*

Artificial Neural Networks. New York, NY: Springer-Verlag, pp. 127–162.

van Rijsbergen, C. J (1979). Information Retrieval. 2nd ed. London, UK: Buttersworth.

West Publishing Company. (1993). WESTLAW Reference Manual. 5th ed. St. Paul, MN: West Publishing Co.

Whittaker, E. T., Watson, G. N. (1990). A Course in Modern Analysis. 4th ed.. Cambridge, UK: Cambridge University Press.

Wilson, P. (1973). Situational relevance. *Information Storage and Retrieval* 9(7):457–471.

Xu, L., Krzyzak, A., Suen, C. Y. (1992). Methods of combining classifiers and their applications to handwriting recognition. *IEEE Transactions on Systems, Man, and Cybernetics* 22(3):418–435.

Index